Water Pollution and Treatment

Water Pollution and Treatment

Edited by
Ezra Duncan

Larsen & Keller
www.larsen-keller.com

Water Pollution and Treatment
Edited by Ezra Duncan
ISBN: 978-1-63549-290-3 (Hardback)

☰ Larsen & Keller

Published by Larsen and Keller Education,
5 Penn Plaza,
19th Floor,
New York, NY 10001, USA

Cataloging-in-Publication Data

Water pollution and treatment / edited by Ezra Duncan.
    p. cm.
Includes bibliographical references and index.
ISBN 978-1-63549-290-3
1. Water--Pollution. 2. Water--Purification. 3. Sewage--Purification. 4. Water quality.
5. Waterborne infection. 6. Water--Pollution--Toxicology. I. Duncan, Ezra.
TD420 .W38 2017
628.168--dc23

The publisher's policy is to use permanent paper from mills that operate a sustainable forestry policy. Furthermore, the publisher ensures that the text paper and cover boards used have met acceptable environmental accreditation standards.

Printed and bound in the United States of America.

For more information regarding Larsen and Keller Education and its products, please visit the publisher's website www.larsen-keller.com

Table of Contents

Preface   **VII**

Chapter 1 **Introduction to Water Pollution**   **1**

Chapter 2 **Types of Water Pollution**   **15**
- Thermal Pollution   15
- Marine Pollution   18
- Groundwater Pollution   29
- Drug Pollution   38
- Nutrient Pollution   39

Chapter 3 **Causes of Water Pollution**   **43**
- Marine Debris   43
- Chemical Waste   51
- Acid Mine Drainage   56

Chapter 4 **Water Quality Parameters Model**   **64**
- Hydrological Transport Model   64
- Groundwater Model   68
- DSSAM Model   74
- Storm Water Management Model   76

Chapter 5 **Various Wastewater Treatments**   **95**
- Wastewater Treatment   95
- Industrial Wastewater Treatment   100
- Agricultural Wastewater Treatment   109
- Sewage Treatment   115
- Reclaimed Water   131

Chapter 6 **Water Purification and its Methods**   **145**
- Water Purification   145
- Filtration   161
- Sedimentation (Water Treatment)   166
- Distillation   170
- Water Chlorination   187

Chapter 7 **Waterborne Diseases**   **192**
- Amoebiasis   192
- Cryptosporidiosis   198
- Schistosomiasis   208
- Dracunculiasis   216
- Enterobiasis   225

Chapter 8   **Aquatic Toxicology: An Overview**   231
               • Aquatic Toxicology   231

Chapter 9   **Laws Relating to Water Safety and Quality Management**   238
               • Safe Drinking Water Act   238
               • Clean Water Act   243

               Permissions

               Index

Preface

This book presents the basic concepts of water pollution in detail. It provides thorough insights into the various concepts and theories and practices used to properly utilize water and control its pollution. Water pollution is one of the critical concepts we are facing today. It refers to the contamination of water bodies like lakes, rivers, oceans, seas, etc. because of industrial waste, paints, grease, oil spills and other harmful chemicals. This textbook is an elaborate overview of the fundamental concepts of water pollution and its treatment. The topics that have been presented are of utmost significance and are bound to provide incredible insights to students. The text will serve as a reference to a broad spectrum of readers.

A short introduction to every chapter is written below to provide an overview of the content of the book:

Chapter 1 - The contamination of water bodies is water pollution, and this form of environmental degradation occurs when contaminants are directly or indirectly discharged into the water body. The chapter on water pollution offers an insightful focus, keeping in mind the complex subject matter; **Chapter 2** - Water pollution can best be understood with the major topics listed in the following chapter. The major categories dealt within this chapter are thermal pollution, marine pollution, groundwater pollution, drug pollution and nutrient pollution. When harmful or potentially harmful chemicals, industrial or residential waste are disposed into the ocean, it causes marine pollution while drug pollution is the pollution of the environment with pharmaceutical drugs which reaches the marine environment through wastewater; **Chapter 3** - The major causes of water pollution are discussed in this chapter. The causes discussed are marine debris, chemical waste and acid mine drainage. The waste created by humans, that has been released in a lake, sea, or ocean is called marine debris, while the outflow of acidic water from metal mines or coal mines is acid mine drainage. The following text will not only provide an overview, it will also delve into the topics related to it; **Chapter 4** - Water quality modeling commits to the prediction of water pollution using mathematical simulation techniques. A typical water quality model is a collection of formulations representing physical mechanisms that determine the position of pollutants in a water body. The chapter strategically encompasses and incorporates the major components of water quality modeling; **Chapter 5** - Wastewater treatment is the process of converting wastewater into useable water. It is water which has been harmfully affected by human influence. The various treatments discussed in the following content are industrial wastewater treatment, agricultural wastewater treatment, sewage treatment and reclaimed water. The chapter strategically encompasses and incorporates the major components and key concepts of water pollution, providing a complete understanding; **Chapter 6** - The process of removing undesirable chemicals or biological contaminants from contaminated water is water purification. Most of the water purified is purified for human consumption, but there can be other purposes as well. The methods of water purification explained in the following script are filtration, sedimentation and water chlorination. The major components of water purification are discussed in this chapter; **Chapter 7 -** Pathogenic microorganisms that are most commonly transmitted by contaminated fresh water cause waterborne diseases. The waterborne diseases that are dealt within this chapter are amoebiasism, cryptosporidiosis, schistosomiasis and enterobiasis. This chapter is a compilation of the various branches of water pollution and management that form an integral part of the broader subject matter; **Chapter 8** – The study of the harm done to the aquatic organisms by humans is known as aquatic toxicology. The fields concerned with aquatic toxicology are freshwater, marine water and sediment environments. Aquatic toxicology is an emerging field of study, the following chapter will not only provide an overview but will also delve into the topics related to it; **Chapter 9** - The plan to ensure the safety of drinking water through the use of a comprehensive risk assessment and risk management approach is a water safety plan. The Safe Drinking Water Act (SDWA) is a law that is intended to ensure safe drinking water for the public. The topics discussed in the chapter are of great importance to broaden the existing knowledge on water safety and quality management.

I extend my sincere thanks to the publisher for considering me worthy of this task. Finally, I thank my family for being a source of support and help.

Editor

Introduction to Water Pollution

The contamination of water bodies is water pollution, and this form of environmental degradation occurs when contaminants are directly or indirectly discharged into the water body. The chapter on water pollution offers an insightful focus, keeping in mind the complex subject matter.

Water pollution is the contamination of water bodies (e.g. lakes, rivers, oceans, aquifers and groundwater). This form of environmental degradation occurs when pollutants are directly or indirectly discharged into water bodies without adequate treatment to remove harmful compounds.

Raw sewage and industrial waste in the New River as it passes from Mexicali to Calexico, California

Water pollution affects the entire biosphere – plants and organisms living in these bodies of water. In almost all cases the effect is damaging not only to individual species and population, but also to the natural biological communities.

Introduction

Water pollution is a major global problem which requires ongoing evaluation and revision of water resource policy at all levels (international down to individual aquifers and wells). It has been suggested that water pollution is the leading worldwide cause of deaths and diseases, and that it accounts for the deaths of more than 14,000 people daily. An estimated 580 people in India die of water pollution related illness every day. About 90 percent of the water in the cities of China is polluted. As of 2007, half a billion Chinese had no access to safe drinking water. In addition to the acute problems of water pollution in developing countries, developed countries also continue to struggle with pollution problems. For example, in the most recent national report on water quality in the United States, 44 percent of assessed stream miles, 64 percent of assessed lake acres, and 30 percent of assessed bays and estuarine square miles were classified as polluted. The head of China's national development agency said in 2007 that one quarter the length of China's seven main rivers were so poisoned the water harmed the skin.

Pollution in the Lachine Canal, Canada

Water is typically referred to as polluted when it is impaired by anthropogenic contaminants and either does not support a human use, such as drinking water, or undergoes a marked shift in its ability to support its constituent biotic communities, such as fish. Natural phenomena such as volcanoes, algae blooms, storms, and earthquakes also cause major changes in water quality and the ecological status of water.

Categories

Although interrelated, surface water and groundwater have often been studied and managed as separate resources. Surface water seeps through the soil and becomes groundwater. Conversely, groundwater can also feed surface water sources. Sources of surface water pollution are generally grouped into two categories based on their origin.

Point Sources

Point source pollution – Shipyard – Rio de Janeiro.

Point source water pollution refers to contaminants that enter a waterway from a single, identifiable source, such as a pipe or ditch. Examples of sources in this category include discharges from a sewage treatment plant, a factory, or a city storm drain. The U.S. Clean Water Act (CWA) defines point source for regulatory enforcement purposes. The CWA definition of point source was amended in 1987 to include municipal storm sewer systems, as well as industrial storm water, such as from construction sites.

Non-point Sources

Nonpoint source pollution refers to diffuse contamination that does not originate from a single discrete source. NPS pollution is often the cumulative effect of small amounts of contaminants gathered from a large area. A common example is the leaching out of nitrogen compounds from fertilized agricultural lands. Nutrient runoff in storm water from "sheet flow" over an agricultural field or a forest are also cited as examples of NPS pollution.

Blue drain and yellow fish symbol used by the UK Environment Agency to raise awareness of the ecological impacts of contaminating surface drainage

Contaminated storm water washed off of parking lots, roads and highways, called urban runoff, is sometimes included under the category of NPS pollution. However, because this runoff is typically channeled into storm drain systems and discharged through pipes to local surface waters, it becomes a point source.

Groundwater Pollution

Interactions between groundwater and surface water are complex. Consequently, groundwater pollution, also referred to as groundwater contamination, is not as easily classified as surface water pollution. By its very nature, groundwater aquifers are susceptible to contamination from sources that may not directly affect surface water bodies, and the distinction of point vs. non-point source may be irrelevant. A spill or ongoing release of chemical or radionuclide contaminants into soil (located away from a surface water body) may not create point or non-point source pollution but can contaminate the aquifer below, creating a toxic plume. The movement of the plume, called a plume front, may be analyzed through a hydrological transport model or groundwater model. Analysis of groundwater contamination may focus on soil characteristics and site geology, hydrogeology, hydrology, and the nature of the contaminants.

Causes

The specific contaminants leading to pollution in water include a wide spectrum of chemicals, pathogens, and physical changes such as elevated temperature and discoloration. While many of the chemicals and substances that are regulated may be naturally occurring (calcium, sodium, iron, manganese, etc.) the concentration is often the key in determining what is a natural component of water and what is a contaminant. High concentrations of naturally occurring substances can have negative impacts on aquatic flora and fauna.

Oxygen-depleting substances may be natural materials such as plant matter (e.g. leaves and grass) as well as man-made chemicals. Other natural and anthropogenic substances may cause turbidity (cloudiness) which blocks light and disrupts plant growth, and clogs the gills of some fish species.

Many of the chemical substances are toxic. Pathogens can produce waterborne diseases in either human or animal hosts. Alteration of water's physical chemistry includes acidity (change in pH), electrical conductivity, temperature, and eutrophication. Eutrophication is an increase in the concentration of chemical nutrients in an ecosystem to an extent that increases in the primary productivity of the ecosystem. Depending on the degree of eutrophication, subsequent negative environmental effects such as anoxia (oxygen depletion) and severe reductions in water quality may occur, affecting fish and other animal populations.

Pathogens

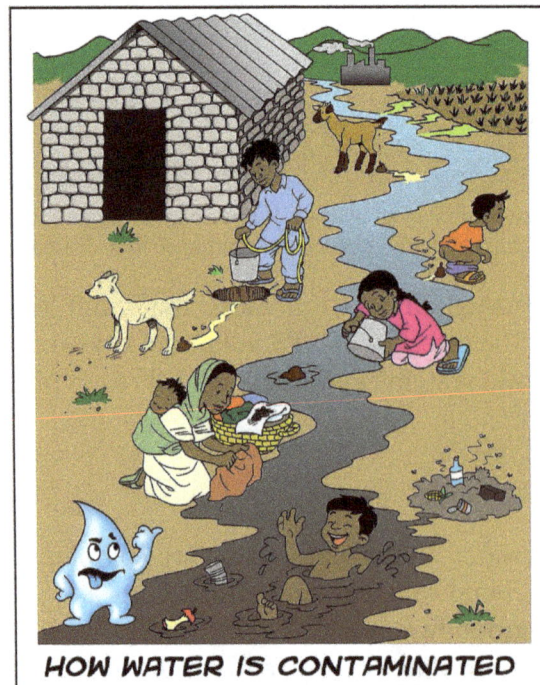

Poster to teach people in South Asia about human activities leading to the pollution of water sources

Disease-causing microorganisms are referred to as pathogens. Although the vast majority of bacteria are either harmless or beneficial, a few pathogenic bacteria can cause disease. Coliform bacteria, which are not an actual cause of disease, are commonly used as a bacterial indicator of water pollution. Other microorganisms sometimes found in surface waters that have caused human health problems include:

- *Burkholderia pseudomallei*
- *Cryptosporidium parvum*
- *Giardia lamblia*
- *Salmonella*
- *Norovirus* and other viruses
- *Parasitic worms including the Schistosoma type*

A manhole cover unable to contain a sanitary sewer overflow.

High levels of pathogens may result from on-site sanitation systems (septic tanks, pit latrines) or inadequately treated sewage discharges. This can be caused by a sewage plant designed with less than secondary treatment (more typical in less-developed countries). In developed countries, older cities with aging infrastructure may have leaky sewage collection systems (pipes, pumps, valves), which can cause sanitary sewer overflows. Some cities also have combined sewers, which may discharge untreated sewage during rain storms.

Muddy river polluted by sediment.

Pathogen discharges may also be caused by poorly managed livestock operations.

Organic, Inorganic and Macroscopic Contaminants

Contaminants may include organic and inorganic substances.

A garbage collection boom in an urban-area stream in Auckland, New Zealand.

Organic water pollutants include:

- Detergents

- Disinfection by-products found in chemically disinfected drinking water, such as chloroform

- Food processing waste, which can include oxygen-demanding substances, fats and grease

- Insecticides and herbicides, a huge range of organohalides and other chemical compounds

- Petroleum hydrocarbons, including fuels (gasoline, diesel fuel, jet fuels, and fuel oil) and lubricants (motor oil), and fuel combustion byproducts, from storm water runoff

- Volatile organic compounds, such as industrial solvents, from improper storage.

- Chlorinated solvents, which are dense non-aqueous phase liquids, may fall to the bottom of reservoirs, since they don't mix well with water and are denser.

    o Polychlorinated biphenyl (PCBs)

    o Trichloroethylene

- Perchlorate

- Various chemical compounds found in personal hygiene and cosmetic products

- Drug pollution involving pharmaceutical drugs and their metabolites

Inorganic water pollutants include:

- Acidity caused by industrial discharges (especially sulfur dioxide from power plants)

- Ammonia from food processing waste

- Chemical waste as industrial by-products

- Fertilizers containing nutrients--nitrates and phosphates—which are found in storm water runoff from agriculture, as well as commercial and residential use

- Heavy metals from motor vehicles (via urban storm water runoff) and acid mine drainage

- Silt (sediment) in runoff from construction sites, logging, slash and burn practices or land clearing sites.

Macroscopic pollution – large visible items polluting the water – may be termed "floatables" in an urban storm water context, or marine debris when found on the open seas, and can include such items as:

- Trash or garbage (e.g. paper, plastic, or food waste) discarded by people on the ground, along with accidental or intentional dumping of rubbish, that are washed by rainfall into storm drains and eventually discharged into surface waters

- Nurdles, small ubiquitous waterborne plastic pellets

- Shipwrecks, large derelict ships.

The Brayton Point Power Station in Massachusetts discharges heated water to Mount Hope Bay.

Thermal Pollution

Thermal pollution is the rise or fall in the temperature of a natural body of water caused by human influence. Thermal pollution, unlike chemical pollution, results in a change in the physical properties of water. A common cause of thermal pollution is the use of water as a coolant by power plants and industrial manufacturers. Elevated water temperatures decrease oxygen levels, which can kill fish and alter food chain composition, reduce species biodiversity, and foster invasion by new thermophilic species. Urban runoff may also elevate temperature in surface waters.

Thermal pollution can also be caused by the release of very cold water from the base of reservoirs into warmer rivers.

Transport and Chemical Reactions of Water Pollutants

Most water pollutants are eventually carried by rivers into the oceans. In some areas of the world the influence can be traced one hundred miles from the mouth by studies using hydrology transport models. Advanced computer models such as SWMM or the DSSAM Model have been used in many locations worldwide to examine the fate of pollutants in aquatic systems. Indicator filter feeding species such as copepods have also been used to study pollutant fates in the New York Bight, for example. The highest toxin loads are not directly at the mouth of the Hudson River, but 100 km (62 mi) south, since several days are required for incorporation into planktonic tissue. The Hudson discharge flows south along the coast due to the coriolis force. Further south are areas of oxygen depletion caused by chemicals using up oxygen and by algae blooms, caused by excess nutrients from algal cell death and decomposition. Fish and shellfish kills have been reported, because toxins climb the food chain after small fish consume copepods, then large fish eat smaller fish, etc. Each successive step up the food chain causes a cumulative concentration of pollutants such as heavy metals (e.g. mercury) and persistent organic pollutants such as DDT. This is known as bio-magnification, which is occasionally used interchangeably with bio-accumulation.

A polluted river draining an abandoned copper mine on Anglesey

Large gyres (vortexes) in the oceans trap floating plastic debris. The North Pacific Gyre, for example, has collected the so-called "Great Pacific Garbage Patch", which is now estimated to be one hundred times the size of Texas. Plastic debris can absorb toxic chemicals from ocean pollution, potentially poisoning any creature that eats it. Many of these long-lasting pieces wind up in the stomachs of marine birds and animals. This results in obstruction of digestive pathways, which leads to reduced appetite or even starvation.

Many chemicals undergo reactive decay or chemical change, especially over long periods of time in groundwater reservoirs. A noteworthy class of such chemicals is the chlorinated hydrocarbons such as trichloroethylene (used in industrial metal degreasing and electronics manufacturing) and tetrachloroethylene used in the dry cleaning industry. Both of these chemicals, which are carcinogens themselves, undergo partial decomposition reactions, leading to new hazardous chemicals (including dichloroethylene and vinyl chloride).

Groundwater pollution is much more difficult to abate than surface pollution because groundwater can move great distances through unseen aquifers. Non-porous aquifers such as clays partially purify water of bacteria by simple filtration (adsorption and absorption), dilution, and, in some cases, chemical reactions and biological activity; however, in some cases, the pollutants merely transform to soil contaminants. Groundwater that moves through open fractures and caverns is not filtered and can be transported as easily as surface water. In fact, this can be aggravated by the human tendency to use natural sinkholes as dumps in areas of karst topography.

There are a variety of secondary effects stemming not from the original pollutant, but a derivative condition. An example is silt-bearing surface runoff, which can inhibit the penetration of sunlight through the water column, hampering photosynthesis in aquatic plants.

Measurement

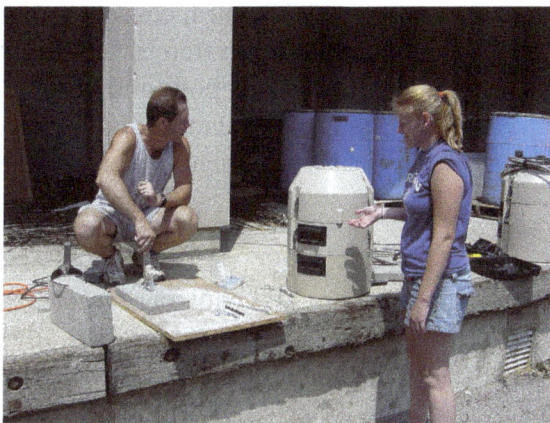

Environmental scientists preparing water autosamplers.

Water pollution may be analyzed through several broad categories of methods: physical, chemical and biological. Most involve collection of samples, followed by specialized analytical tests. Some methods may be conducted *in situ*, without sampling, such as temperature. Government agencies and research organizations have published standardized, validated analytical test methods to facilitate the comparability of results from disparate testing events.

Sampling

Sampling of water for physical or chemical testing can be done by several methods, depending on the accuracy needed and the characteristics of the contaminant. Many contamination events are sharply restricted in time, most commonly in association with rain events. For this reason "grab" samples are often inadequate for fully quantifying contaminant levels. Scientists gathering this type of data often employ auto-sampler devices that pump increments of water at either time or discharge intervals.

Sampling for biological testing involves collection of plants and/or animals from the surface water body. Depending on the type of assessment, the organisms may be identified for biosurveys (population counts) and returned to the water body, or they may be dissected for bioassays to determine toxicity.

Physical Testing

Common physical tests of water include temperature, solids concentrations (e.g., total suspended solids (TSS)) and turbidity.

Chemical Testing

Water samples may be examined using the principles of analytical chemistry. Many published test methods are available for both organic and inorganic compounds. Frequently used methods include pH, biochemical oxygen demand (BOD), chemical oxygen demand (COD), nutrients (nitrate and phosphorus compounds), metals (including copper, zinc, cadmium, lead and mercury), oil and grease, total petroleum hydrocarbons (TPH), and pesticides.

Biological Testing

Biological testing involves the use of plant, animal, and/or microbial indicators to monitor the health of an aquatic ecosystem. They are any biological species or group of species whose function, population, or status can reveal what degree of ecosystem or environmental integrity is present. One example of a group of bio-indicators are the copepods and other small water crustaceans that are present in many water bodies. Such organisms can be monitored for changes (biochemical, physiological, or behavioral) that may indicate a problem within their ecosystem.

Control of Pollution

Decisions on the type and degree of treatment and control of wastes, and the disposal and use of adequately treated wastewater, must be based on a consideration all the technical factors of each drainage basin, in order to prevent any further contamination or harm to the environment.

Sewage Treatment

In urban areas of developed countries, domestic sewage is typically treated by centralized sewage treatment plants. Well-designed and operated systems (i.e., secondary treatment or better) can remove 90 percent or more of the pollutant load in sewage. Some plants have additional systems to remove nutrients and pathogens.

Deer Island Wastewater Treatment Plant serving Boston, Massachusetts and vicinity.

Cities with sanitary sewer overflows or combined sewer overflows employ one or more engineering approaches to reduce discharges of untreated sewage, including:

- utilizing a green infrastructure approach to improve storm water management capacity throughout the system, and reduce the hydraulic overloading of the treatment plant

- repair and replacement of leaking and malfunctioning equipment

- increasing overall hydraulic capacity of the sewage collection system (often a very expensive option).

A household or business not served by a municipal treatment plant may have an individual septic tank, which pre-treats the wastewater on site and infiltrates it into the soil.

Industrial Wastewater Treatment

Some industrial facilities generate ordinary domestic sewage that can be treated by municipal facilities. Industries that generate wastewater with high concentrations of conventional pollutants (e.g. oil and grease), toxic pollutants (e.g. heavy metals, volatile organic compounds) or other non-conventional pollutants such as ammonia, need specialized treatment systems. Some of these facilities can install a pre-treatment system to remove the toxic components, and then send the partially treated wastewater to the municipal system. Industries generating large volumes of wastewater typically operate their own complete on-site treatment systems. Some industries have been successful at redesigning their manufacturing processes to reduce or eliminate pollutants, through a process called pollution prevention.

Heated water generated by power plants or manufacturing plants may be controlled with:

- cooling ponds, man-made bodies of water designed for cooling by evaporation, convection, and radiation

- cooling towers, which transfer waste heat to the atmosphere through evaporation and/or heat transfer

- cogeneration, a process where waste heat is recycled for domestic and/or industrial heating purposes.

Riparian buffer lining a creek in Iowa.

Agricultural Wastewater Treatment

Non Point Source Controls

Sediment (loose soil) washed off fields is the largest source of agricultural pollution in the United States. Farmers may utilize erosion controls to reduce runoff flows and retain soil on their fields. Common techniques include contour plowing, crop mulching, crop rotation, planting perennial crops and installing riparian buffers.

Nutrients (nitrogen and phosphorus) are typically applied to farmland as commercial fertilizer, animal manure, or spraying of municipal or industrial wastewater (effluent) or sludge. Nutrients may also enter runoff from crop residues, irrigation water, wildlife, and atmospheric deposition. Farmers can develop and implement nutrient management plans to reduce excess application of nutrients and reduce the potential for nutrient pollution.

To minimize pesticide impacts, farmers may use Integrated Pest Management (IPM) techniques (which can include biological pest control) to maintain control over pests, reduce reliance on chemical pesticides, and protect water quality.

Feedlot in the United States

Point Source Wastewater Treatment

Farms with large livestock and poultry operations, such as factory farms, are called *concentrated animal feeding operations* or *feedlots* in the US and are being subject to increasing government regulation. Animal slurries are usually treated by containment in anaerobic lagoons before disposal by spray or trickle application to grassland. Constructed wetlands are sometimes used to facilitate treatment of animal wastes. Some animal slurries are treated by mixing with straw and composted at high temperature to produce a bacteriologically sterile and friable manure for soil improvement.

Erosion and Sediment Control from Construction Sites

Sediment from construction sites is managed by installation of:

- erosion controls, such as mulching and hydroseeding, and

- sediment controls, such as sediment basins and silt fences.

Silt fence installed on a construction site.

Discharge of toxic chemicals such as motor fuels and concrete washout is prevented by use of:

- spill prevention and control plans, and

- specially designed containers (e.g. for concrete washout) and structures such as overflow controls and diversion berms.

Control of Urban Runoff (Storm Water)

Effective control of urban runoff involves reducing the velocity and flow of storm water, as well as reducing pollutant discharges. Local governments use a variety of storm water management techniques to reduce the effects of urban runoff. These techniques, called best management practices (BMPs) in the U.S., may focus on water quantity control, while others focus on improving water quality, and some perform both functions.

Retention basin for controlling urban runoff

Pollution prevention practices include low-impact development techniques, installation of green roofs and improved chemical handling (e.g. management of motor fuels & oil, fertilizers and pesticides). Runoff mitigation systems include infiltration basins, bioretention systems, constructed wetlands, retention basins and similar devices.

Thermal pollution from runoff can be controlled by storm water management facilities that absorb the runoff or direct it into groundwater, such as bioretention systems and infiltration basins. Retention basins tend to be less effective at reducing temperature, as the water may be heated by the sun before being discharged to a receiving stream.

References

- "Low Impact Development and Other Green Design Strategies". National Pollutant Discharge Elimination System. EPA. 2014. Archived from the original on 2015-02-19.

- Wachman, Richard (2007-12-09). "Water becomes the new oil as world runs dry". The Guardian. London. Retrieved 2015-09-23.

- U.S. Natural Resources Conservation Service (NRCS). Washington, DC. "National Conservation Practice Standards." National Handbook of Conservation Practices. Accessed 2015-10-02.

- "An overview of diarrhea, symptoms, diagnosis and the costs of morbidity" (PDF). CHNRI. 2010. Archived from the original (PDF) on May 12, 2013.

- For example, Rodger B.; Clesceri, Leonore S.; Eaton, Andrew D.; et al., eds. (2012). Standard Methods for the Examination of Water and Wastewater (22nd ed.). Washington, DC: American Public Health Association. ISBN 978-0875530130. .

2

Types of Water Pollution

Water pollution can best be understood with the major topics listed in the following chapter. The major categories dealt within this chapter are thermal pollution, marine pollution, groundwater pollution, drug pollution and nutrient pollution. When harmful or potentially harmful chemicals, industrial or residential waste are disposed into the ocean, it causes marine pollution while drug pollution is the pollution of the environment with pharmaceutical drugs which reaches the marine environment through wastewater.

Thermal Pollution

Thermal pollution is the degradation of water quality by any process that changes ambient water temperature. A common cause of thermal pollution is the use of water as a coolant by power plants and industrial manufacturers. When water used as a coolant is returned to the natural environment at a higher temperature, the change in temperature decreases oxygen supply and affects ecosystem composition. Fish and other organisms adapted to particular temperature range can be killed by an abrupt change in water temperature (either a rapid increase or decrease) known as "thermal shock."

Urban runoff—stormwater discharged to surface waters from roads and parking lots—can also be a source of elevated water temperatures.

Ecological Effects

Potrero Generating Station discharged heated water into San Francisco Bay. The plant was closed in 2011.

Warm Water

Elevated temperature typically decreases the level of dissolved oxygen of water. This can harm aquatic animals such as fish, amphibians and other aquatic organisms. Thermal pollution may also increase the metabolic rate of aquatic animals, as enzyme activity, resulting in these organisms consuming more food in a shorter time than if their environment were not changed. An increased metabolic rate may result in fewer resources; the more adapted organisms moving in may have an advantage over organisms that are not used to the warmer temperature. As a result, food chains of the old and new environments may be compromised. Some fish species will avoid stream segments or coastal areas adjacent to a thermal discharge. Biodiversity can be decreased as a result.

High temperature limits oxygen dispersion into deeper waters, contributing to anaerobic conditions. This can lead to increased bacteria levels when there is ample food supply. Many aquatic species will fail to reproduce at elevated temperatures.

Primary producers (e.g. plants, cyanobacteria) are affected by warm water because higher water temperature increases plant growth rates, resulting in a shorter lifespan and species overpopulation. This can cause an algae bloom which reduces oxygen levels.

Temperature changes of even one to two degrees Celsius can cause significant changes in organism metabolism and other adverse cellular biology effects. Principal adverse changes can include rendering cell walls less permeable to necessary osmosis, coagulation of cell proteins, and alteration of enzyme metabolism. These cellular level effects can adversely affect mortality and reproduction.

A large increase in temperature can lead to the denaturing of life-supporting enzymes by breaking down hydrogen- and disulphide bonds within the quaternary structure of the enzymes. Decreased enzyme activity in aquatic organisms can cause problems such as the inability to break down lipids, which leads to malnutrition.

In limited cases, warm water has little deleterious effect and may even lead to improved function of the receiving aquatic ecosystem. This phenomenon is seen especially in seasonal waters and is known as thermal enrichment. An extreme case is derived from the aggregational habits of the manatee, which often uses power plant discharge sites during winter. Projections suggest that manatee populations would decline upon the removal of these discharges.

Cold Water

Releases of unnaturally cold water from reservoirs can dramatically change the fish and macroinvertebrate fauna of rivers, and reduce river productivity. In Australia, where many rivers have warmer temperature regimes, native fish species have been eliminated, and macroinvertebrate fauna have been drastically altered. This may be mitigated by designing the dam to release warmer surface waters instead of the colder water at the bottom of the reservoir.

Thermal Shock

When a power plant first opens or shuts down for repair or other causes, fish and other organisms adapted to particular temperature range can be killed by the abrupt change in water temperature, either an increase or decrease, known as "thermal shock."

Sources and Control of Thermal Pollution

Cooling tower at Gustav Knepper Power Station, Dortmund, Germany

Industrial Wastewater

In the United States, about 75 to 82 percent of thermal pollution is generated by power plants. The remainder is from industrial sources such as petroleum refineries, pulp and paper mills, chemical plants, steel mills and smelters. Heated water from these sources may be controlled with:

- cooling ponds, man-made bodies of water designed for cooling by evaporation, convection, and radiation

- cooling towers, which transfer waste heat to the atmosphere through evaporation and/or heat transfer

- cogeneration, a process where waste heat is recycled for domestic and/or industrial heating purposes.

Some facilities use once-through cooling (OTC) systems which do not reduce temperature as effectively as the above systems. For example, the Potrero Generating Station in San Francisco (closed in 2011), used OTC and discharged water to San Francisco Bay approximately 10 °C (20 °F) above the ambient bay temperature.

A bioretention cell for treating urban runoff in California

Urban Runoff

During warm weather, urban runoff can have significant thermal impacts on small streams, as stormwater passes over hot parking lots, roads and sidewalks. Stormwater management facilities that absorb runoff or direct it into groundwater, such as bioretention systems and infiltration basins, can reduce these thermal effects. These and related systems for managing runoff are components of an expanding urban design approach commonly called green infrastructure.

Retention basins (stormwater ponds) tend to be less effective at reducing runoff temperature, as the water may be heated by the sun before being discharged to a receiving stream.

Marine Pollution

While marine pollution can be obvious, as with the marine debris shown above, it is often the pollutants that cannot be seen that cause most harm.

Marine pollution occurs when harmful, or potentially harmful, effects result from the entry into the ocean of chemicals, particles, industrial, agricultural and residential waste, noise, or the spread of invasive organisms. Eighty percent of marine pollution comes from land. Air pollution is also a contributing factor by carrying off pesticides or dirt into the ocean. Land and air pollution have proven to be harmful to marine life and its habitats.

The pollution often comes from non point sources such as agricultural runoff, wind-blown debris and dust. Nutrient pollution, a form of water pollution, refers to contamination by excessive inputs of nutrients. It is a primary cause of eutrophication of surface waters, in which excess nutrients, usually nitrogen or phosphorus, stimulate algae growth.

Many potentially toxic chemicals adhere to tiny particles which are then taken up by plankton and benthos animals, most of which are either deposit or filter feeders. In this way, the toxins are concentrated upward within ocean food chains. Many particles combine chemically in a manner highly depletive of oxygen, causing estuaries to become anoxic.

When pesticides are incorporated into the marine ecosystem, they quickly become absorbed into marine food webs. Once in the food webs, these pesticides can cause mutations, as well as diseases, which can be harmful to humans as well as the entire food web.

Toxic metals can also be introduced into marine food webs. These can cause a change to tissue matter, biochemistry, behaviour, reproduction, and suppress growth in marine life. Also, many animal feeds have a high fish meal or fish hydrolysate content. In this way, marine toxins can be transferred to land animals, and appear later in meat and dairy products.

History

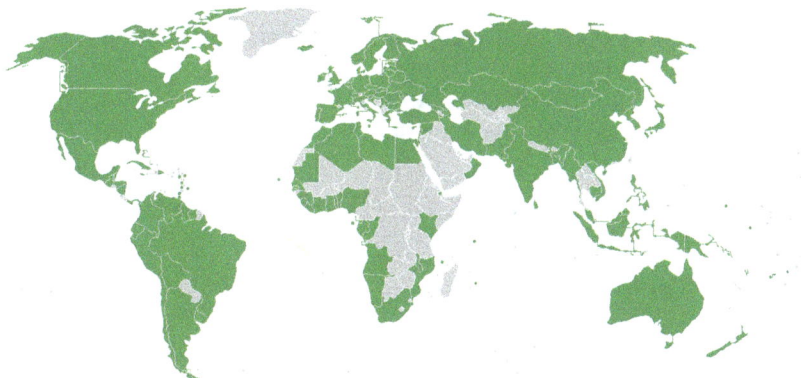

Parties to the MARPOL 73/78 convention on marine pollution

Although marine pollution has a long history, significant international laws to counter it were only enacted in the twentieth century. Marine pollution was a concern during several United Nations Conferences on the Law of the Sea beginning in the 1950s. Most scientists believed that the oceans were so vast that they had unlimited ability to dilute, and thus render pollution harmless.

In the late 1950s and early 1960s, there were several controversies about dumping radioactive waste off the coasts of the United States by companies licensed by the Atomic Energy Commission, into the Irish Sea from the British reprocessing facility at Windscale, and into the Mediterranean Sea by the French Commissariat à l'Energie Atomique. After the Mediterranean Sea controversy, for example, Jacques Cousteau became a worldwide figure in the campaign to stop marine pollution. Marine pollution made further international headlines after the 1967 crash of the oil tanker Torrey Canyon, and after the 1969 Santa Barbara oil spill off the coast of California.

Marine pollution was a major area of discussion during the 1972 United Nations Conference on the Human Environment, held in Stockholm. That year also saw the signing of the Convention on the Prevention of Marine Pollution by Dumping of Wastes and Other Matter, sometimes called the London Convention. The London Convention did not ban marine pollution, but it established black and gray lists for substances to be banned (black) or regulated by national authorities (gray). Cyanide and high-level radioactive waste, for example, were put on the black list. The London Convention applied only to waste dumped from ships, and thus did nothing to regulate waste discharged as liquids from pipelines.

Pathways of Pollution

There are many different ways to categorize, and examine the inputs of pollution into our marine ecosystems. Patin (n.d.) notes that generally there are three main types of inputs of pollution into the ocean: direct discharge of waste into the oceans, runoff into the waters due to rain, and pollutants that are released from the atmosphere.

One common path of entry by contaminants to the sea are rivers. The evaporation of water from oceans exceeds precipitation. The balance is restored by rain over the continents entering rivers and then being returned to the sea. The Hudson in New York State and the Raritan in New Jersey, which empty at the northern and southern ends of Staten Island, are a source of mercury contamination of zooplankton (copepods) in the open ocean. The highest concentration in the filter-feeding copepods is not at the mouths of these rivers but 70 miles south, nearer Atlantic City, because water flows close to the coast. It takes a few days before toxins are taken up by the plankton.

Pollution is often classed as point source or nonpoint source pollution. Point source pollution occurs when there is a single, identifiable, and localized source of the pollution. An example is directly discharging sewage and industrial waste into the ocean. Pollution such as this occurs particularly in developing nations. Nonpoint source pollution occurs when the pollution comes from ill-defined and diffuse sources. These can be difficult to regulate. Agricultural runoff and wind blown debris are prime examples.

Direct Discharge

Pollutants enter rivers and the sea directly from urban sewerage and industrial waste discharges, sometimes in the form of hazardous and toxic wastes.

Inland mining for copper, gold. etc., is another source of marine pollution. Most of the pollution is simply soil, which ends up in rivers flowing to the sea. However, some minerals discharged in the course of the mining can cause problems, such as copper, a common industrial pollutant, which can interfere with the life history and development of coral polyps. Mining has a poor environmental track record. For example, according to the United States Environmental Protection Agency, mining has contaminated portions of the headwaters of over 40% of watersheds in the western continental US. Much of this pollution finishes up in the sea.

Land Runoff

Surface runoff from farming, as well as urban runoff and runoff from the construction of roads, buildings, ports, channels, and harbours, can carry soil and particles laden with carbon, nitrogen, phosphorus, and minerals. This nutrient-rich water can cause fleshy algae and phytoplankton to thrive in coastal areas; known as algal blooms, which have the potential to create hypoxic conditions by using all available oxygen.

Polluted runoff from roads and highways can be a significant source of water pollution in coastal areas. About 75% of the toxic chemicals that flow into Puget Sound are carried by stormwater that runs off paved roads and driveways, rooftops, yards and other developed land.

Ship Pollution

A cargo ship pumps ballast water over the side.

Ships can pollute waterways and oceans in many ways. Oil spills can have devastating effects. While being toxic to marine life, polycyclic aromatic hydrocarbons (PAHs), found in crude oil, are very difficult to clean up, and last for years in the sediment and marine environment.

Oil spills are probably the most emotive of marine pollution events. However, while a tanker wreck may result in extensive newspaper headlines, much of the oil in the world's seas comes from other smaller sources, such as tankers discharging ballast water from oil tanks used on return ships, leaking pipelines or engine oil disposed of down sewers.

Discharge of cargo residues from bulk carriers can pollute ports, waterways and oceans. In many instances vessels intentionally discharge illegal wastes despite foreign and domestic regulation prohibiting such actions. It has been estimated that container ships lose over 10,000 containers at sea each year (usually during storms). Ships also create noise pollution that disturbs natural wildlife, and water from ballast tanks can spread harmful algae and other invasive species.

Ballast water taken up at sea and released in port is a major source of unwanted exotic marine life. The invasive freshwater zebra mussels, native to the Black, Caspian and Azov seas, were probably transported to the Great Lakes via ballast water from a transoceanic vessel. Meinesz believes that one of the worst cases of a single invasive species causing harm to an ecosystem can be attributed to a seemingly harmless jellyfish. *Mnemiopsis leidyi*, a species of comb jellyfish that spread so it now inhabits estuaries in many parts of the world. It was first introduced in 1982, and thought to have been transported to the Black Sea in a ship's ballast water. The population of the jellyfish shot up exponentially and, by 1988, it was wreaking havoc upon the local fishing industry. "The anchovy catch fell from 204,000 tons in 1984 to 200 tons in 1993; sprat from 24,600 tons in 1984 to 12,000 tons in 1993; horse mackerel from 4,000 tons in 1984 to zero in 1993." Now that the jellyfish have exhausted the zooplankton, including fish larvae, their numbers have fallen dramatically, yet they continue to maintain a stranglehold on the ecosystem.

Invasive species can take over once occupied areas, facilitate the spread of new diseases, introduce new genetic material, alter underwater seascapes and jeopardize the ability of native species to obtain food. Invasive species are responsible for about $138 billion annually in lost revenue and management costs in the US alone.

Atmospheric Pollution

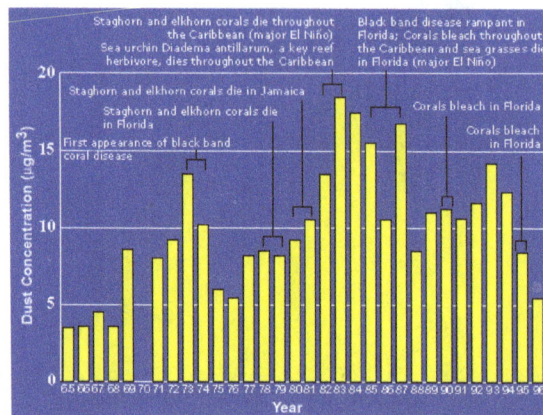

Graph linking atmospheric dust to various coral deaths across the Caribbean Sea and Florida

Another pathway of pollution occurs through the atmosphere. Wind blown dust and debris, including plastic bags, are blown seaward from landfills and other areas. Dust from the Sahara moving around the southern periphery of the subtropical ridge moves into the Caribbean and Florida during the warm season as the ridge builds and moves northward through the subtropical Atlantic. Dust can also be attributed to a global transport from the Gobi and Taklamakan deserts across Korea, Japan, and the Northern Pacific to the Hawaiian Islands. Since 1970, dust outbreaks have worsened due to periods of drought in Africa. There is a large variability in dust transport to the Caribbean and Florida from year to year; however, the flux is greater during positive phases of the North Atlantic Oscillation. The USGS links dust events to a decline in the health of coral reefs across the Caribbean and Florida, primarily since the 1970s.

Climate change is raising ocean temperatures and raising levels of carbon dioxide in the atmosphere. These rising levels of carbon dioxide are acidifying the oceans. This, in turn, is altering aquatic ecosystems and modifying fish distributions, with impacts on the sustainability of fisheries and the livelihoods of the communities that depend on them. Healthy ocean ecosystems are also important for the mitigation of climate change.

Deep Sea Mining

Deep sea mining is a relatively new mineral retrieval process that takes place on the ocean floor. Ocean mining sites are usually around large areas of polymetallic nodules or active and extinct hydrothermal vents at about 1,400 – 3,700 meters below the ocean's surface. The vents create sulfide deposits, which contain precious metals such as silver, gold, copper, manganese, cobalt, and zinc. The deposits are mined using either hydraulic pumps or bucket systems that take ore to the surface to be processed. As with all mining operations, deep sea mining raises questions about environmental damages to the surrounding areas

Because deep sea mining is a relatively new field, the complete consequences of full scale mining operations are unknown. However, experts are certain that removal of parts of the sea floor will result in disturbances to the benthic layer, increased toxicity of the water column and sediment plumes from tailings. Removing parts of the sea floor disturbs the habitat of benthic organisms, possibly, depending on the type of mining and location, causing permanent disturbances. Aside from direct impact of mining the area, leakage, spills and corrosion would alter the mining area's chemical makeup.

Among the impacts of deep sea mining, sediment plumes could have the greatest impact. Plumes are caused when the tailings from mining (usually fine particles) are dumped back into the ocean, creating a cloud of particles floating in the water. Two types of plumes occur: near bottom plumes and surface plumes. Near bottom plumes occur when the tailings are pumped back down to the mining site. The floating particles increase the turbidity, or cloudiness, of the water, clogging filter-feeding apparatuses used by benthic organisms. Surface plumes cause a more serious problem. Depending on the size of the particles and water currents the plumes could spread over vast areas. The plumes could impact zooplankton and light penetration, in turn affecting the food web of the area.

Types of Pollution

Acidification

The oceans are normally a natural carbon sink, absorbing carbon dioxide from the atmosphere. Because the levels of atmospheric carbon dioxide are increasing, the oceans are becoming more acidic. The potential consequences of ocean acidification are not fully understood, but there are concerns that structures made of calcium carbonate may become vulnerable to dissolution, affecting corals and the ability of shellfish to form shells.

Island with fringing reef in the Maldives. Coral reefs are dying around the world.

Oceans and coastal ecosystems play an important role in the global carbon cycle and have removed about 25% of the carbon dioxide emitted by human activities between 2000 and 2007 and about half the anthropogenic CO_2 released since the start of the industrial revolution. Rising ocean temperatures and ocean acidification means that the capacity of the ocean carbon sink will gradually get weaker, giving rise to global concerns expressed in the Monaco and Manado Declarations.

A report from NOAA scientists published in the journal Science in May 2008 found that large amounts of relatively acidified water are upwelling to within four miles of the Pacific continental shelf area of North America. This area is a critical zone where most local marine life lives or is born. While the paper dealt only with areas from Vancouver to northern California, other continental shelf areas may be experiencing similar effects.

A related issue is the methane clathrate reservoirs found under sediments on the ocean floors. These trap large amounts of the greenhouse gas methane, which ocean warming has the potential to release. In 2004 the global inventory of ocean methane clathrates was estimated to occupy between one and five million cubic kilometres. If all these clathrates were to be spread uniformly across the ocean floor, this would translate to a thickness between three and fourteen metres. This estimate corresponds to 500–2500 gigatonnes carbon (Gt C), and can be compared with the 5000 Gt C estimated for all other fossil fuel reserves.

Eutrophication

Eutrophication is an increase in chemical nutrients, typically compounds containing nitrogen or phosphorus, in an ecosystem. It can result in an increase in the ecosystem's primary productivity (excessive plant growth and decay), and further effects including lack of oxygen and severe reductions in water quality, fish, and other animal populations.

Polluted lagoon.

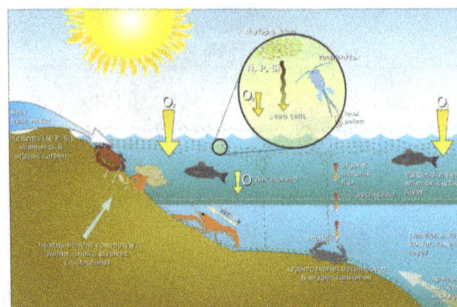

Effect of eutrophication on marine benthic life

The biggest culprit are rivers that empty into the ocean, and with it the many chemicals used as fertilizers in agriculture as well as waste from livestock and humans. An excess of oxygen depleting chemicals in the water can lead to hypoxia and the creation of a dead zone.

Estuaries tend to be naturally eutrophic because land-derived nutrients are concentrated where runoff enters the marine environment in a confined channel. The World Resources Institute has identified 375 hypoxic coastal zones around the world, concentrated in coastal areas in Western Europe, the Eastern and Southern coasts of the US, and East Asia, particularly in Japan. In the ocean, there are frequent red tide algae blooms that kill fish and marine mammals and cause respiratory problems in humans and some domestic animals when the blooms reach close to shore.

In addition to land runoff, atmospheric anthropogenic fixed nitrogen can enter the open ocean. A study in 2008 found that this could account for around one third of the ocean's external (non-recycled) nitrogen supply and up to three per cent of the annual new marine biological production. It has been suggested that accumulating reactive nitrogen in the environment may have consequences as serious as putting carbon dioxide in the atmosphere.

One proposed solution to eutrophication in estuaries is to restore shellfish populations, such as oysters. Oyster reefs remove nitrogen from the water column and filter out suspended solids, subsequently reducing the likelihood or extent of harmful algal blooms or anoxic conditions. Filter feeding activity is considered beneficial to water quality by controlling phytoplankton density and sequestering nutrients, which can be removed from the system through shellfish harvest, buried in the sediments, or lost through denitrification. Foundational work toward the idea of improving marine water quality through shellfish cultivation to was conducted by Odd Lindahl et al., using mussels in Sweden.

Plastic Debris

Marine debris is mainly discarded human rubbish which floats on, or is suspended in the ocean. Eighty percent of marine debris is plastic – a component that has been rapidly accumulating since the end of World War II. The mass of plastic in the oceans may be as high as 100,000,000 tonnes (98,000,000 long tons; 110,000,000 short tons).

A mute swan builds a nest using plastic garbage.

Discarded plastic bags, six pack rings and other forms of plastic waste which finish up in the ocean present dangers to wildlife and fisheries. Aquatic life can be threatened through entanglement, suffocation, and ingestion. Fishing nets, usually made of plastic, can be left or lost in the ocean by fishermen. Known as ghost nets, these entangle fish, dolphins, sea turtles, sharks, dugongs, crocodiles, seabirds, crabs, and other creatures, restricting movement, causing starvation, laceration and infection, and, in those that need to return to the surface to breathe, suffocation.

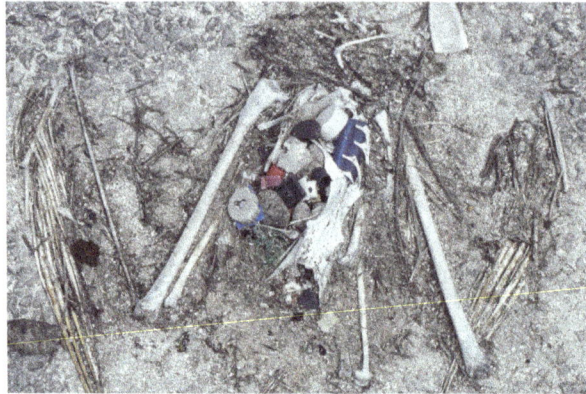

Remains of an albatross containing ingested flotsam

Many animals that live on or in the sea consume flotsam by mistake, as it often looks similar to their natural prey. Plastic debris, when bulky or tangled, is difficult to pass, and may become permanently lodged in the digestive tracts of these animals. Especially when evolutionary adaptions make it impossible for the likes of turtles to reject plastic bags, which resemble jellyfish when immersed in water, as they have a system in their throat to stop slippery foods from otherwise escaping. Thereby blocking the passage of food and causing death through starvation or infection.

Plastics accumulate because they don't biodegrade in the way many other substances do. They will photodegrade on exposure to the sun, but they do so properly only under dry conditions, and water inhibits this process. In marine environments, photodegraded plastic disintegrates into ever smaller pieces while remaining polymers, even down to the molecular level. When floating plastic particles photodegrade down to zooplankton sizes, jellyfish attempt to consume them, and in this way the plastic enters the ocean food chain. Many of these long-lasting pieces end up in the stomachs of marine birds and animals, including sea turtles, and black-footed albatross.

Marine debris on Kamilo Beach, Hawaii, washed up from the Great Pacific Garbage Patch

Plastic debris tends to accumulate at the centre of ocean gyres. In particular, the Great Pacific Garbage Patch has a very high level of plastic particulate suspended in the upper water column.

In samples taken in 1999, the mass of plastic exceeded that of zooplankton (the dominant animal life in the area) by a factor of six. Midway Atoll, in common with all the Hawaiian Islands, receives substantial amounts of debris from the garbage patch. Ninety percent plastic, this debris accumulates on the beaches of Midway where it becomes a hazard to the bird population of the island. Midway Atoll is home to two-thirds (1.5 million) of the global population of Laysan albatross. Nearly all of these albatross have plastic in their digestive system and one-third of their chicks die.

Toxic additives used in the manufacture of plastic materials can leach out into their surroundings when exposed to water. Waterborne hydrophobic pollutants collect and magnify on the surface of plastic debris, thus making plastic far more deadly in the ocean than it would be on land. Hydrophobic contaminants are also known to bioaccumulate in fatty tissues, biomagnifying up the food chain and putting pressure on apex predators. Some plastic additives are known to disrupt the endocrine system when consumed, others can suppress the immune system or decrease reproductive rates. Floating debris can also absorb persistent organic pollutants from seawater, including PCBs, DDT and PAHs. Aside from toxic effects, when ingested some of these are mistaken by the animal brain for estradiol, causing hormone disruption in the affected wildlife.

Toxins

Apart from plastics, there are particular problems with other toxins that do not disintegrate rapidly in the marine environment. Examples of persistent toxins are PCBs, DDT, TBT, pesticides, furans, dioxins, phenols and radioactive waste. Heavy metals are metallic chemical elements that have a relatively high density and are toxic or poisonous at low concentrations. Examples are mercury, lead, nickel, arsenic and cadmium. Such toxins can accumulate in the tissues of many species of aquatic life in a process called bioaccumulation. They are also known to accumulate in benthic environments, such as estuaries and bay muds: a geological record of human activities of the last century.

Specific examples

- Chinese and Russian industrial pollution such as phenols and heavy metals in the Amur River have devastated fish stocks and damaged its estuary soil.

- Wabamun Lake in Alberta, Canada, once the best whitefish lake in the area, now has unacceptable levels of heavy metals in its sediment and fish.

- Acute and chronic pollution events have been shown to impact southern California kelp forests, though the intensity of the impact seems to depend on both the nature of the contaminants and duration of exposure.

- Due to their high position in the food chain and the subsequent accumulation of heavy metals from their diet, mercury levels can be high in larger species such as bluefin and albacore. As a result, in March 2004 the United States FDA issued guidelines recommending that pregnant women, nursing mothers and children limit their intake of tuna and other types of predatory fish.

- Some shellfish and crabs can survive polluted environments, accumulating heavy metals or toxins in their tissues. For example, mitten crabs have a remarkable ability to survive in highly modified aquatic habitats, including polluted waters. The farming and harvesting of such species needs careful management if they are to be used as a food.

- Surface runoff of pesticides can alter the gender of fish species genetically, transforming male into female fish.

- Heavy metals enter the environment through oil spills – such as the Prestige oil spill on the Galician coast – or from other natural or anthropogenic sources.

- In 2005, the 'Ndrangheta, an Italian mafia syndicate, was accused of sinking at least 30 ships loaded with toxic waste, much of it radioactive. This has led to widespread investigations into radioactive-waste disposal rackets.

- Since the end of World War II, various nations, including the Soviet Union, the United Kingdom, the United States, and Germany, have disposed of chemical weapons in the Baltic Sea, raising concerns of environmental contamination.

Underwater Noise

Marine life can be susceptible to noise or the sound pollution from sources such as passing ships, oil exploration seismic surveys, and naval low-frequency active sonar. Sound travels more rapidly and over larger distances in the sea than in the atmosphere. Marine animals, such as cetaceans, often have weak eyesight, and live in a world largely defined by acoustic information. This applies also to many deeper sea fish, who live in a world of darkness. Between 1950 and 1975, ambient noise at one location in the Pacific Ocean increased by about ten decibels (that is a tenfold increase in intensity).

Noise also makes species communicate louder, which is called the Lombard vocal response. Whale songs are longer when submarine-detectors are on. If creatures don't "speak" loud enough, their voice can be masked by anthropogenic sounds. These unheard voices might be warnings, finding of prey, or preparations of net-bubbling. When one species begins speaking louder, it will mask other species voices, causing the whole ecosystem to eventually speak louder.

According to the oceanographer Sylvia Earle, "Undersea noise pollution is like the death of a thousand cuts. Each sound in itself may not be a matter of critical concern, but taken all together, the noise from shipping, seismic surveys, and military activity is creating a totally different environment than existed even 50 years ago. That high level of noise is bound to have a hard, sweeping impact on life in the sea."

Adaptation and Mitigation

Much anthropogenic pollution ends up in the ocean. The 2011 edition of the United Nations Environment Programme Year Book identifies as the main emerging environmental issues the loss to the oceans of massive amounts of phosphorus, "a valuable fertilizer needed to feed a growing global population", and the impact billions of pieces of plastic waste are having globally on the health of marine environments. Bjorn Jennssen (2003) notes in his article, "Anthropogenic pollution may reduce biodiversity and productivity of marine ecosystems, resulting in reduction and depletion of human marine food resources". There are two ways the overall level of this pollution can be mitigated: either the human population is reduced, or a way is found to reduce the ecological footprint left behind by the average human. If the second way is not adopted, then the first way may be imposed as world ecosystems falter.

Aerosol can polluting a beach.

The second way is for humans, individually, to pollute less. That requires social and political will, together with a shift in awareness so more people respect the environment and are less disposed to abuse it. At an operational level, regulations, and international government participation is needed. It is often very difficult to regulate marine pollution because pollution spreads over international barriers, thus making regulations hard to create as well as enforce.

Without appropriate awareness of marine pollution, the necessary global will to effectively address the issues may prove inadequate. Balanced information on the sources and harmful effects of marine pollution need to become part of general public awareness, and ongoing research is required to fully establish, and keep current, the scope of the issues. As expressed in Daoji and Dag's research, one of the reasons why environmental concern is lacking among the Chinese is because the public awareness is low and therefore should be targeted. Likewise, regulation, based upon such in-depth research should be employed. In California, such regulations have already been put in place to protect Californian coastal waters from agricultural runoff. This includes the California Water Code, as well as several voluntary programs. Similarly, in India, several tactics have been employed that help reduce marine pollution, however, they do not significantly target the problem. In Chennai, sewage has been dumped further into open waters. Due to the mass of waste being deposited, open-ocean is best for diluting, and dispersing pollutants, thus making them less harmful to marine ecosystems.

Groundwater Pollution

Groundwater pollution (also called groundwater contamination) occurs when pollutants are released to the ground and make their way down into groundwater. It can also occur naturally due to the presence of a minor and unwanted constituent, contaminant or impurity in the groundwater, in which case it is more likely referred to as contamination rather than pollution.

Groundwater pollution example in Lusaka, Zambia where the pit latrine in the background is polluting the shallow well in the foreground with pathogens and nitrate.

The pollutant creates a contaminant plume within an aquifer. Movement of water and dispersion within the aquifer spreads the pollutant over a wider area. Its advancing boundary, often called a plume edge, can intersect with groundwater wells or daylight into surface water such as seeps and spring, making the water supplies unsafe for humans and wildlife. The movement of the plume, called a plume front, may be analyzed through a hydrological transport model or groundwater model. Analysis of groundwater pollution may focus on soil characteristics and site geology, hydrogeology, hydrology, and the nature of the contaminants.

Pollution can occur from on-site sanitation systems, landfills, effluent from wastewater treatment plants, leaking sewers, petrol stations or from over application of fertilizers in agriculture. Pollution (or contamination) can also occur from naturally occurring contaminants, such as arsenic or fluoride. Using polluted groundwater causes hazards to public health through poisoning or the spread of disease.

Different mechanisms have influence on the transport of pollutants, e.g. diffusion, adsorption, precipitation, decay, in the groundwater. The interaction of groundwater contamination with surface waters is analyzed by use of hydrology transport models.

Pollutant Types

Contaminants found in groundwater cover a broad range of physical, inorganic chemical, organic chemical, bacteriological, and radioactive parameters. Principally, many of the same pollutants that play a role in surface water pollution may also be found in polluted groundwater, although their respective importance may differ.

Pathogens

Pathogens contained in feces can lead to groundwater pollution when they are given the opportunity to reach the groundwater, making it unsafe for drinking. Of the four pathogen types that are present in feces (bacteria, viruses, protozoa and helminths or helminth eggs), the first three can be commonly found in polluted groundwater, whereas the relatively large helminth eggs are usually filtered out by the soil matrix.

Waterborne diseases can be spread via a groundwater well which is contaminated with fecal pathogens from pit latrines

Groundwater that is contaminated with pathogens can lead to fatal fecal-oral transmission of diseases (e.g. cholera, diarrhoea).

If the local hydrogeological conditions (which can vary within a space of a few square kilometres) are ignored, pit latrines can cause significant public health risks via contaminated groundwater.

Nitrate

In addition to the issue of pathogens, there is also the issue of nitrate pollution in groundwater from pit latrines, which has led to numerous cases of "blue baby syndrome" in children, notably in rural countries such as Romania and Bulgaria. Nitrate levels above 10 mg/L (10 ppm) in groundwater can cause "blue baby syndrome" (acquired methemoglobinemia).

Nitrate can also enter the groundwater via excessive use of fertilizers, including manure. This is because only a fraction of the nitrogen-based fertilizers is converted to produce and other plant matter. The remainder accumulates in the soil or lost as run-off. High application rates of nitrogen-containing fertilizers combined with the high water-solubility of nitrate leads to increased runoff into surface water as well as leaching into groundwater, thereby causing groundwater pollution. The excessive use of nitrogen-containing fertilizers (be they synthetic or natural) is particularly damaging, as much of the nitrogen that is not taken up by plants is transformed into nitrate which is easily leached.

The nutrients, especially nitrates, in fertilizers can cause problems for natural habitats and for human health if they are washed off soil into watercourses or leached through soil into groundwater.

Volatile Organic Compounds

Volatile organic compounds (VOCs) are a dangerous contaminant of groundwater. They are generally introduced to the environment through careless industrial practices. Many of these compounds were not known to be harmful until the late 1960s and it was some time before regular testing of groundwater identified these substances in drinking water sources.

Others

Organic pollutants can also be found in groundwater, such as insecticides and herbicides, a range of organohalides and other chemical compounds, petroleum hydrocarbons, various chemical compounds found in personal hygiene and cosmetic products, drug pollution involving pharmaceutical drugs and their metabolites. Inorganic pollutans might include ammonia, nitrate, phosphate, heavy metals or radionuclides.

Naturally Occurring

Arsenic

In the Ganges Plain of northern India and Bangladesh severe contamination of groundwater by naturally occurring arsenic affects 25% of water wells in the shallower of two regional aquifers. The pollution occurs because aquifer sediments contain organic matter that generates anaerobic conditions in the aquifer. These conditions result in the microbial dissolution of iron oxides in the sediment and, thus, the release of the arsenic, normally strongly bound to iron oxides, into the water. As a consequence, arsenic-rich groundwater is often iron-rich, although secondary processes often obscure the association of dissolved arsenic and dissolved iron.

Fluoride

In areas that have naturally occurring high levels of fluoride in groundwater which is used for drinking water, both dental and skeletal fluorosis can be prevalent and severe.

Causes

Landfill Leachate

Leachate from sanitary landfills can lead to groundwater pollution.

Love Canal was one of the most widely known examples of groundwater pollution. In 1978, residents of the Love Canal neighborhood in upstate New York noticed high rates of cancer and an alarming number of birth defects. This was eventually traced to organic solvents and dioxins from an industrial landfill that the neighborhood had been built over and around, which had then infiltrated into the water supply and evaporated in basements to further contaminate the air. Eight hundred families were reimbursed for their homes and moved, after extensive legal battles and media coverage.

On-site Sanitation Systems

Groundwater pollution with pathogens and nitrate can also occur from the liquids infiltrating into the ground from on-site sanitation systems such as pit latrines and septic tanks, depending on the population density and the hydrogeological conditions.

Liquids leach from the pit and pass the unsaturated soil zone (which is not completely filled with water). Subsequently, these liquids from the pit enter the groundwater where they may lead to groundwater pollution. This is a problem if a nearby water well is used to supply groundwater for drinking water purposes. During the passage in the soil, pathogens can die off or be adsorbed significantly, mostly depending on the travel time between the pit and the well. Most, but not all pathogens die within 50 days of travel through the subsurface.

A traditional housing compound near Herat, Afghanistan, where a shallow water supply well (foreground) is in close proximity to the pit latrine (behind the white greenhouse) leading to contamination of the groundwater.

The degree of pathogen removal strongly varies with soil type, aquifer type, distance and other environmental factors. For this reason, it is difficult to estimate the safe distance between a pit latrine or a septic tank and a water source. In any case, such recommendations about the safe distance are mostly ignored by those building pit latrines. In addition, household plots are of a limited size and therefore pit latrines are often built much closer to groundwater wells than what can be regarded as safe. This results in groundwater pollution and household members falling sick when using this groundwater as a source of drinking water.

Sewage Treatment Plants

The treated effluent from sewage treatment plants may also reach the aquifer if the effluent is infiltrated or discharged to local surface water bodies. Therefore, those substances that are not removed in conventional sewage treatment plants may reach the groundwater as well.

For example, detected concentrations of pharmaceutical residues in groundwater were in the order of 50 ng/L in several locations in Germany. This is because in conventional sewage treatment plants, micro-pollutants such as hormones, pharmaceutical residues and other micro-pollutants contained in urine and feces are only partially removed and the remainder is discharged into surface water, from where it may also reach the groundwater.

Hydraulic Fracturing

The recent growth of Hydraulic Fracturing ("Fracking") wells in the United States has raised valid concerns regarding its potential risks of contaminating groundwater resources. The Environmental Protection Agency (EPA), along with many other researchers, has been delegated to study the relationship between hydraulic fracturing and drinking water resources. While the EPA has not found significant evidence of a widespread, systematic impact on drinking water by hydraulic fracturing, this may be due to insufficient systematic pre- and post- hydraulic fracturing data on drinking water quality, and the presence of other agents of contamination that preclude the link between shale oil/gas extraction and its impact.

Despite the EPA's lack of profound widespread evidence, other researchers have made significant observations of rising groundwater contamination in close proximity to major shale oil/gas drilling sites located in Marcellus (Northeastern Pennsylvania) and Horn River Basins (British Columbia, Canada). Within one kilometer of these specific sites, a subset of shallow drinking water consistently showed higher concentration levels of methane, ethane, and propane concentrations than normal. An evaluation of higher Helium and other noble gas concentration along with the rise of hydrocarbon levels supports the distinction between hydraulic fracturing fugitive gas and naturally occurring "background" hydrocarbon content. This contamination is speculated to be the result of leaky, failing, or improperly installed gas well casings. Furthermore, it is theorized that contamination could also result from the capillary migration of deep residual hyper-saline water and hydraulic fracturing fluid, slowly flowing through faults and fractures until finally making contact with groundwater resources; however, many researchers argue that the permeability of rocks overlying shale formations are too low to allow this to ever happen sufficiently. To ultimately prove this theory, there would have to be traces of toxic trihalomethanes (THM) since they are often associated with the presence of stray gas contamination, and typically co-occur with high halogen concentrations in hyper-saline waters.

While conclusions regarding groundwater pollution as the result to hydraulic fracturing fluid flow is restricted in both space and time, researchers have hypothesized that the potential for systematic stray gas contamination depends mainly on the integrity of the shale oil/gas well structure, along with its relative geological location to local fracture systems that could potentially provide flow paths for fugitive gas migration.

Though widespread, systematic contamination by hydraulic fracturing has been heavily disputed, one major source of contamination that has the most consensus among researchers of being the most problematic is site-specific accidental spillage of hydraulic fracturing fluid and produced water. So far, a significant majority of groundwater contamination events are derived from surface-level anthropogenic routes rather than the subsurface flow from underlying shale formations. Examples of such events include: a fracking fluid spillage in Acorn Fork Creek, Kentucky that caused a widespread death among aquatic species in 2007; a 420,000 gallon spillage of hyper-saline produced water that turned a once very-fertile farmland in New Mexico into a dead-zone in 2010; and a 42,000 gallon fracking fluid spillage in Arlington, Texas that necessitated an evacuation of over a 100 homes in 2015. While the damage can be obvious, and much more effort is being done to prevent these accidents from occurring so frequently, the lack of data from fracking oil spills continue to leave researchers in the dark. In many of these events, the data acquired from the leakage or spillage is often very vague, and thus would lead researchers to lacking conclusions.

Other

Further causes of groundwater pollution are excessive application of fertilizer or pesticides, chemical spills from commercial or industrial operations, chemical spills occurring during transport (e.g. spillage of diesel fuels), illegal waste dumping, infiltration from urban runoff or mining operations, road salts, de-icing chemicals from airports and even atmospheric contaminants since groundwater is part of the hydrologic cycle. Over application of animal manure may also result in groundwater pollution with pharmaceutical residues.

Groundwater pollution can also occur from leaking sewers which has been observed for example in Germany.

Mechanisms

The passage of water through the subsurface can provide a reliable natural barrier to contamination but it only works under favorable conditions.

The stratigraphy of the area plays an important role in the transport of pollutants. An area can have layers of sandy soil, fractured bedrock, clay, or hardpan. Areas of karst topography on limestone bedrock are sometimes vulnerable to surface pollution from groundwater. Earthquake faults can also be entry routes for downward contaminant entry. Water table conditions are of great importance for drinking water supplies, agricultural irrigation, waste disposal (including nuclear waste), wildlife habitat, and other ecological issues.

Interactions with Surface Water

Although interrelated, surface water and groundwater have often been studied and managed as separate resources. Surface water seeps through the soil and becomes groundwater. Conversely, groundwater can also feed surface water sources. Sources of surface water pollution are generally grouped into two categories based on their origin.

Interactions between groundwater and surface water are complex. Consequently, groundwater pollution, sometimes referred to as groundwater contamination, is not as easily classified as surface water pollution. By its very nature, groundwater aquifers are susceptible to contamination from sources that may not directly affect surface water bodies, and the distinction of point vs. non-point source may be irrelevant. A spill or ongoing release of chemical or radionuclide contaminants into soil (located away from a surface water body) may not create point or non-point source pollution but can contaminate the aquifer below, creating a toxic plume.

Prevention

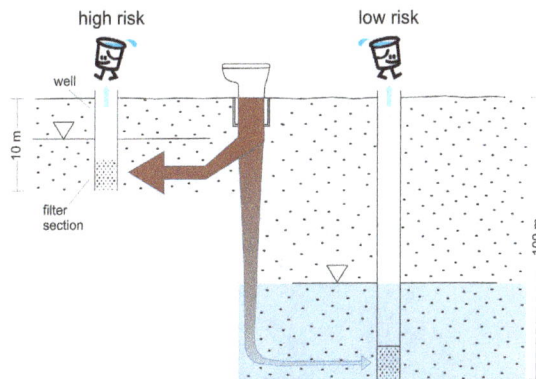

Schematic showing that there is a lower risk of groundwater pollution with greater depth of the water well

Locating on-site Sanitation Systems

On-site sanitation systems can be designed in such a way that groundwater pollution from these sanitation systems is prevented from occurring. Detailed guidelines have been developed to estimate safe distances to protect groundwater sources from pollution from on-site sanitation. The following criteria have been proposed for safe siting (i.e. deciding on the location) of on-site sanitation systems:

- Horizontal distance between the drinking water source and the sanitation system

    o Guideline values for horizontal separation distances between on-site sanitation systems and water sources vary widely (e.g. 15 to 100 m horizontal distance between pit latrine and groundwater wells)

- Vertical distance between drinking water well and sanitation system
- Aquifer type
- Groundwater flow direction
- Impermeable layers
- Slope and surface drainage
- Volume of leaking wastewater
- Superposition, i.e. the need to consider a larger planning area

As a very general guideline it is recommended that the bottom of the pit should be at least 2 m above groundwater level, and a minimum horizontal distance of 30 m between a pit and a water source is normally recommended to limit exposure to microbial contamination. However, no general statement should be made regarding the minimum lateral separation distances required to prevent contamination of a well from a pit latrine. For example, even 50 m lateral separation distance might not be sufficient in a strongly karstified system with a downgradient supply well or spring, while 10 m lateral separation distance is completely sufficient if there is a well developed clay cover layer and the annular space of the groundwater well is well sealed.

Legislation

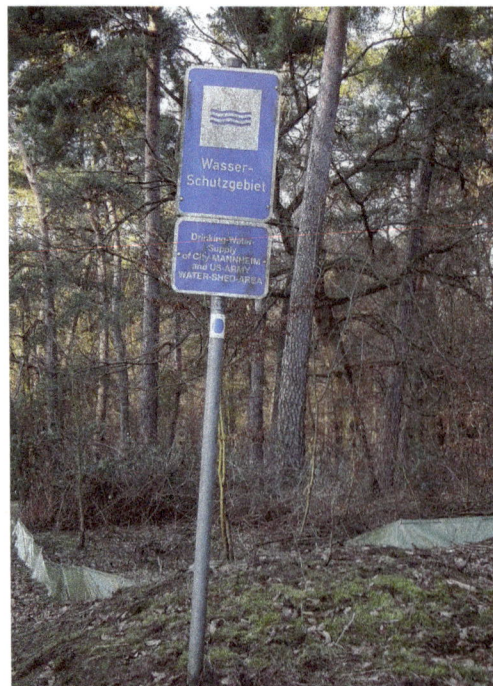

Sign near Mannheim, Germany indicating a zone as a dedicated "groundwater protection zone"

United States

In November 2006, the Environmental Protection Agency published the Ground Water Rule in the United States Federal Register. The EPA was worried that the ground water system would be vulnerable to contamination from fecal matter. The point of the rule was to keep microbial pathogens out of public water sources. The 2006 Ground Water Rule was an amendment of the 1996 Safe Drinking Water Act.

The ways to deal with groundwater pollution that has already occurred can be grouped into the following categories: containing the pollutants to prevent them from migrating further; removing the pollutants from the aquifer; remediating the aquifer by either immobilizing or detoxifying the contaminants while they are still in the aquifer (in-situ); treating the groundwater at its point of use; or abandoning the use of this aquifer's groundwater and finding an alternative source of water.

Management

Point-of-use Treatment

Portable water purification devices or "point-of-use" (POU) water treatment systems and field water disinfection techniques can be used to remove some forms of groundwater pollution prior to drinking, namely any fecal pollution. Many commercial portable water purification systems or chemical additives are available which can remove pathogens, chlorine, bad taste, odors, and heavy metals like lead and mercury.

Techniques include boiling, filtration, activated charcoal absorption, chemical disinfection, ultraviolet purification, ozone water disinfection, solar water disinfection, solar distillation, homemade water filters.

Groundwater Remediation

Groundwater pollution is much more difficult to abate than surface pollution because groundwater can move great distances through unseen aquifers. Non-porous aquifers such as clays partially purify water of bacteria by simple filtration (adsorption and absorption), dilution, and, in some cases, chemical reactions and biological activity; however, in some cases, the pollutants merely transform to soil contaminants. Groundwater that moves through open fractures and caverns is not filtered and can be transported as easily as surface water. In fact, this can be aggravated by the human tendency to use natural sinkholes as dumps in areas of karst topography.

Pollutants and contaminants can be removed from ground water by applying various techniques thereby making it safe for use. Ground water treatment (or remediation) techniques span biological, chemical, and physical treatment technologies. Most ground water treatment techniques utilize a combination of technologies. Some of the biological treatment techniques include bioaugmentation, bioventing, biosparging, bioslurping, and phytoremediation. Some chemical treatment techniques include ozone and oxygen gas injection, chemical precipitation, membrane separation, ion exchange, carbon absorption, aqueous chemical oxidation, and surfactant enhanced recovery. Some chemical techniques may be implemented using nanomaterials. Physical treatment techniques include, but are not limited to, pump and treat, air sparging, and dual phase extraction.

Abandonment

If treatment or remediation of the polluted groundwater is deemed to be too difficult or expensive then abandoning the use of this aquifer's groundwater and finding an alternative source of water is the only other option.

Society and Culture

Hinkley, USA

The town of Hinkley, California (USA), had its groundwater contaminated with hexavalent chromium starting in 1952, resulting in a legal case against Pacific Gas & Electric (PG&E) and a multimillion-dollar settlement in 1996. The legal case was dramatized in the film Erin Brockovich, released in 2000.

Walkerton, Canada

In the year 2000, groundwater pollution occurred in the small town of Walkerton, Canada leading to seven deaths in what is known as the Walkerton E. Coli outbreak. The water supply which was drawn from groundwater became contaminated with the highly dangerous O157:H7 strain of E. coli bacteria. This contamination was due to farm runoff into an adjacent water well that was vulnerable to groundwater pollution.

Lusaka, Zambia

The peri-urban areas of Lusaka, the capital of Zambia, have ground conditions which are strongly karstified and for this reason – together with the increasing population density in these peri-urban areas – pollution of water wells from pit latrines is a major public health threat there.

Drug Pollution

Drug pollution or pharmaceutical pollution is pollution of the environment with pharmaceutical drugs and their metabolites, which reach the marine environment (groundwater, rivers, lakes, and oceans) through wastewater. Drug pollution is therefore mainly a form of water pollution.

"Pharmaceutical pollution is now detected in waters throughout the world," said a scientist at the Cary Institute of Ecosystem Studies in Millbrook, New York. "Causes include aging infrastructure, sewage overflows and agricultural runoff. Even when wastewater makes it to sewage treatment facilities, they aren't equipped to remove pharmaceuticals."

Sources and Effects

Most such pollution comes simply from the drugs having been cleared and excreted in the urine. The portion that comes from expired or unneeded drugs that are flushed unused down the toilet is smaller, but it is also important, especially in hospitals (where its magnitude is greater than in residential contexts). Other sources include agricultural runoff (because of antibiotic use in livestock) and pharmaceutical manufacturing. Drug pollution is implicated in the sex effects of water pollution. It is suspected as a contributor (besides industrial pollution) in fish kills, amphibian dieoffs, and amphibian pathomorphology.

Prevention

The main action for preventing drug pollution is to incinerate unwanted pharmaceutical drugs rather than flushing them down the drain. Burning them chemically degrades their active molecules, with few exceptions. The resulting ash can be further processed before landfilling, such as to remove and recycle any heavy metals that may be present.

There are now programs in many cities that provide collection points at places including drug stores, grocery stores, and police stations. People can bring their unwanted pharmaceuticals there for safe disposal, instead of flushing them (externalizing them to the waterways) or throwing them in the trash (externalizing them to the landfill, where they can become leachate).

Another aspect of drug pollution prevention is environmental law and regulation, although this faces problems of enforcement costs, enforcement corruption and negligence, and, where enforcement succeeds, increased costs of doing business. The lobbying of pros and cons is ongoing.

Manufacturing

One extreme example of drug pollution was found in India in 2009 in an area where pharmaceutical manufacturing activity is concentrated. Not all pharmaceutical manufacturing contributes to the problem. In places where environmental law and regulation are adequately enforced, the wastewater from the factories is cleaned to a safe level. But to the extent that the market rewards "looking the other way" in developing nations, whether through local corruption (bribed inspectors or regulators) or plausible deniability, such protections are circumvented. This problem belongs to everyone, because consumers in well-regulated places constitute the biggest customers of the factories that operate in the inadequately regulated or inspected places, meaning that externality is involved.

Nutrient Pollution

Nutrient pollution caused by Surface runoff of soil and fertilizer during a rain storm

Nutrient pollution, a form of water pollution, refers to contamination by excessive inputs of nutrients. It is a primary cause of eutrophication of surface waters, in which excess nutrients, usually nitrogen or phosphorus, stimulate algal growth. Sources of nutrient pollution include surface run-off from farm fields and pastures, discharges from septic tanks and feedlots, and emissions from combustion. Excess nutrients have been summarized as potentially leading to:

- Population effects: excess growth of algae (blooms);

- Community effects: species composition shifts (dominant taxa);

- Ecological effects:food web changes, light limitation;

- Biogeochemical effects: excess organic carbon (eutrophication); dissolved oxygen deficits (environmental hypoxia); toxin production;

- Human health effects: excess nitrate in drinking water (blue baby syndrome); disinfection by-products in drinking water.

In a 2011 United States Environmental Protection Agency report, the agency's Science Advisory Board succinctly stated: "Excess reactive nitrogen compounds in the environment are associated with many large-scale environmental concerns, including eutrophication of surface waters, toxic algae blooms, hypoxia, acid rain, nitrogen saturation in forests, and global warming."

Excess Nutrients and TMDLs

The regulatory mechanism in the United States, a Total Maximum Daily Load (TMDL), prescribes the maximum amount of a pollutant (including nutrients) that a body of water can receive while still meeting U.S. Clean Water Act water quality standards. Specifically, Section 303 of the Clean Water Act requires each state to generate a TMDL report for each body of water impaired by pollutants. TMDL reports identify pollutant levels and strategies to accomplish pollutant reduction goals. EPA has described TMDLs as establishing a pollutant budget then allocating portions of the overall budget to the pollutant's sources. For many coastal water bodies, the main pollutant issue is excess nutrients, also termed *nutrient over-enrichment*. A TMDL can prescribe the minimum level of Dissolved Oxygen (DO) available in a body of water, which is directly related to nutrient levels. In 2010, 18 percent of TMDLs nationwide were related to nutrient levels including organic enrichment/oxygen depletion, noxious plants, algal growth, and ammonia. In Long Island Sound the TMDL development process enabled the Connecticut Department of Environmental Protection (CTDEP) and the New York State Department of Environmental Conservation (NYSDEC) to incorporate a 58.5 percent nitrogen reduction target into a regulatory and legal framework.

Nutrient Remediation

Innovative solutions have been conceived to deal with nutrient pollution in aquatic systems by altering or enhancing natural processes to shift nutrient effects away from detrimental ecological impacts. Nutrient remediation is a form of environmental remediation, but concerns only biologically active nutrients such as nitrogen and phosphorus. "Remediation" refers to the removal of pollution or contaminants, generally for the protection of human health. In environmental remediation nutrient removal technologies include biofiltration, which uses living material to capture and biologically degrade pollutants. Examples include green belts, riparian areas, natural and constructed wetlands, and treatment ponds. These areas most commonly capture anthropogenic discharges such as wastewater, stormwater runoff, or sewage treatment, for land reclamation after mining, refinery activity, or land development. Biofiltration utilizes biological assimilation to capture, absorb, and eventually incorporate the pollutants (including nutrients) into living tissue. Another form of nutrient removal is bioremediation, which uses microrganisms to remove pollutants. Bioremediation can occur on its own as natural attenuation or intrinsic bioremediation or can be encouraged by the addition of fertilizers, called biostimulation.

Mussels, nutrient bioextractors.

Nutrient bioextraction is the preferred term for bioremediation involving cultured plants and animals. Nutrient bioextraction or bioharvesting is the practice of farming and harvesting shellfish and seaweed for the purpose of removing nitrogen and other nutrients from natural water bodies. It has been suggested that nitrogen removal by oyster reefs could generate net benefits for sources facing nitrogen emission restrictions, similar to other nutrient trading scenarios. Specifically, if oysters maintain nitrogen levels in estuaries below thresholds that would lead to the imposition of emission limits, oysters effectively save the sources the compliance costs they otherwise would incur. Several studies have shown that oysters and mussels have the capacity to dramatically impact nitrogen levels in estuaries.

History of Nutrient Policy in the United States

In 1998, a Policy for a National Nutrient Strategy was created with a focus on developing nutrient criteria. Between 2000-2010 criteria for Rivers/Streams, Lakes/Reservoirs, Estuaries, Wetlands, and guidance were completed, including "ecoregional" nutrient criteria in 14 ecoregions across the U.S. In 2004, the EPA Office of Science and Technology defined EPA's expectations for numeric criteria for total nitrogen (TN), total phosphorus (TP), chlorophyll a(chl-a), and clarity, and established "mutually-agreed upon plans" for state criteria development. In 2007, EPA reiterated EPA's expectations for numeric criteria and committed EPA to support state efforts.

Nutrient Trading

After the EPA had introduced watershed-based NPDES permitting in 2007, interest in nutrient removal and achieving regional TMDLs led to the development of nutrient trading schemes. Nutrient trading is a type of water quality trading, a market-based policy instrument used to improve or maintain water quality. Water quality trading arose around 2005 and is based on the fact that different pollution sources in a watershed can face very different costs to control the same pollutant. Water quality trading involves the voluntary exchange of pollution reduction credits from sources with low costs of pollution control to those with high costs of pollution control, and the same principles apply to nutrient water quality trading. The underlying principle is "polluter pays", usually linked with a regulatory driver for participating is the trading program.

A 2013 Forest Trends report summarized water quality trading programs and found three main types of funders: beneficiaries of watershed protection, polluters compensating for their impacts and 'public good payers' that may not directly benefit, but fund the pollution reduction credits on behalf of a government or NGO. As of 2013, payments were overwhelmingly initiated by public good payers like governments and NGOs.

References

- Kennish, Michael J. (1992). Ecology of Estuaries: Anthropogenic Effects. Marine Science Series. Boca Raton, FL: CRC Press. ISBN 978-0-8493-8041-9.

- Hamblin, Jacob Darwin (2008) Poison in the Well: Radioactive Waste in the Oceans at the Dawn of the Nuclear Age. Rutgers University Press. ISBN 978-0-8135-4220-1

- Warner R (2009) Protecting the oceans beyond national jurisdiction: strengthening the international law framework. Vol. 3 of Legal aspects of sustainable development, Brill, ISBN 978-90-04-17262-3.

- "Brayton Point Station: Final NPDES Permit". NPDES Permits in New England. U.S. Environmental Protection Agency (EPA), Boston, MA. 2014. Retrieved 2015-04-13.

- Administration, US Department of Commerce, National Oceanic and Atmospheric. "What is the biggest source of pollution in the ocean?". oceanservice.noaa.gov. Retrieved 2015-11-22.

- Lynda Knobeloch; Barbara Salna; Adam Hogan; Jeffrey Postle; Henry Anderson (2000). "Blue Babies and Nitrate-Contaminated Well Water". Ehponline.org. Retrieved 23 March 2015.

- "Total Maximum Daily Load (TMDLs) at Work: New York: Restoring the Long Island Sound While Saving Money". EPA. Retrieved June 14, 2013.

- "Midway's albatross population stable | Hawaii's Newspaper". The Honolulu Advertiser. 17 January 2005. Retrieved 20 May 2012.

- C. J. Rosen; B. P. Horgan (9 January 2009). "Preventing Pollution Problems from Lawn and Garden Fertilizers". Extension.umn.edu. Retrieved 25 August 2010.

- "NOFA Interstate Council: The Natural Farmer. Ecologically Sound Nitrogen Management. Mark Schonbeck". Nofa.org. 25 February 2004. Retrieved 25 August 2010.

3

Causes of Water Pollution

The major causes of water pollution are discussed in this chapter. The causes discussed are marine debris, chemical waste and acid mine drainage. The waste created by humans, that has been released in a lake, sea, or ocean is called marine debris, while the outflow of acidic water from metal mines or coal mines is acid mine drainage. The following text will not only provide an overview, it will also delve into the topics related to it.

Marine Debris

Marine debris, also known as marine litter, is human-created waste that has deliberately or accidentally been released in a lake, sea, ocean or waterway. Floating oceanic debris tends to accumulate at the centre of gyres and on coastlines, frequently washing aground, when it is known as *beach litter* or tidewrack. Deliberate disposal of wastes at sea is called *ocean dumping*. Naturally occurring debris, such as driftwood, are also present.

Marine debris on the Hawaiian coast

With the increasing use of plastic, human influence has become an issue as many types of plastics do not biodegrade. Waterborne plastic poses a serious threat to fish, seabirds, marine reptiles, and marine mammals, as well as to boats and coasts. Dumping, container spillages, litter washed into storm drains and waterways and wind-blown landfill waste all contribute to this problem.

Types of Debris

Researchers classify debris as either land- or ocean-based; in 1991, the United Nations Joint Group of Experts on the Scientific Aspects of Marine Pollution estimated that up to 80% of the pollution was land-based. More recent studies have found that more than half of plastic debris found on Korean shores is ocean-based. A wide variety of anthropogenic artifacts can become marine debris; plastic bags, balloons, buoys, rope, medical waste, glass bottles and plastic bottles, cigarette stubs, cigarette lighters, beverage cans, polystyrene, lost fishing line and nets, and various wastes from cruise ships and oil rigs are among the items commonly found to have washed ashore. Six pack rings, in particular, are considered emblematic of the problem. The US military used ocean dumping for unused weapons and bombs, including ordinary bombs, UXO, landmines and chemical weapons from at least 1919 until 1970. Millions of pounds of ordnance were disposed of in the Gulf of Mexico and off the coasts of at least 16 states, from New Jersey to Hawaii (although these, of course, do not wash up onshore, nor is the US the only country who has practiced this, safely dismantling and recycling explosives being a very dangerous and costly procedure).

Debris on beach near Dar es Salaam, Tanzania

Debris collected from beaches on Tern Island in the French Frigate Shoals over one month

Eighty percent of marine debris is plastic. Plastics accumulate because they typically do not biodegrade as many other substances do. They photodegrade on exposure to sunlight, although they do so only under dry conditions, as water inhibits photolysis. In a 2014 study using computers models, scientists from the group 5 Gyres, estimate 5.25 trillion pieces of plastic weighing 269,000 tons dispersed in oceans in similar amount in the Northern and Southern Hemispheres, and one-hundredth of them in particles in the scale of a sand-size.

Ghost Nets

Fishing nets left or lost in the ocean by fishermen – ghost nets – can entangle fish, dolphins, sea turtles, sharks, dugongs, crocodiles, seabirds, crabs and other creatures. These nets restrict movement, causing starvation, laceration and infection, and, in animals that breathe air, suffocation.

Nurdles and Plastic Bags

A handful of nurdles, spilt from a train in Pineville, Louisiana

Nurdles, also known as "mermaids' tears", are plastic pellets, typically under five millimetres in diameter, that are a major component of marine debris. They are a raw material in plastics manufacturing, and enter the natural environment when spilled. Weathering produces ever smaller pieces. Nurdles strongly resemble fish eggs.

Plastic

8.8 million metric tons of plastic waste are dumped in the world's oceans each year. Asia was the leading source of mismanaged plastic waste, with China alone accounting for 2.4 million metric tons.

Plastic waste has reached all the world's oceans. This plastic pollution harms an estimated 100,000 sea turtles and marine mammals and 1,000,000 sea creatures each year. Larger plastics (called macroplastics) such as plastic shopping bags can clog the digestive tracts of these larger animals when consumed by them and can cause starvation through restricting the movement of food, or by filling the stomach and tricking the animal into thinking it is full. Microplastics on the other hand harm smaller marine life. Pelagic plastic pieces in the center of our ocean's gyres for example outnumber live marine plankton, and are passed up the food chain to reach all marine life. A 1994 study of the seabed using trawl nets in the North-Western Mediterranean around the coasts of Spain, France and Italy reported mean concentrations of debris of 1,935 items per square kilometre. Plastic debris accounted for 77%, of which 93% was plastic bags.

Deep-sea Debris

Litter, made from diverse materials that are denser than surface water (such as glasses, metals and some plastics), have been found to spread over the floor of seas and open oceans, where it can become entangled in corals and interfere with other sea-floor life, or even become buried under sediment, making clean-up extremely difficult, especially due to the wide area of its dispersal compared to shipwrecks. Research performed by MBARI found items including plastic bags below 2000m depth off the west coast of North America and around Hawaii.

Sources of Debris

An estimated 10,000 containers at sea each year are lost by container ships, usually during storms. One famous spillage occurred in the Pacific Ocean in 1992, when thousands of rubber ducks and other toys (now known as the "Friendly Floatees") went overboard during a storm. The toys have since been found all over the world, providing a better understanding of ocean currents. Similar incidents have happened before, such as when *Hansa Carrier* dropped 21 containers (with one notably containing buoyant Nike shoes). In 2007, MSC Napoli beached in the English Channel, dropping hundreds of containers, most of which washed up on the Jurassic Coast, a World Heritage Site.

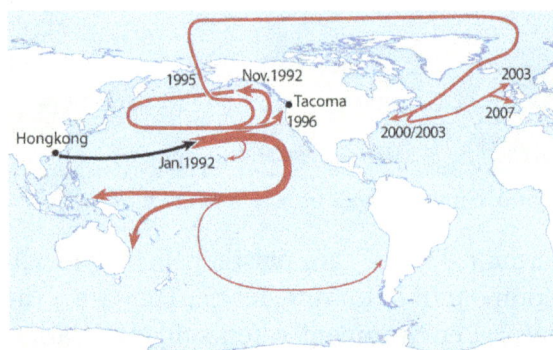

Travel of the Friendly Floatees

In Halifax Harbour, Nova Scotia, 52% of items were generated by recreational use of an urban park, 14% from sewage disposal and only 7% from shipping and fishing activities. Around four fifths of oceanic debris is from rubbish blown onto the water from landfills, and urban runoff. In the 1987 Syringe Tide, medical waste washed ashore in New Jersey after having been blown from Fresh Kills Landfill. On the remote sub-Antarctic island of South Georgia, fishing-related debris, approximately 80% plastics, are responsible for the entanglement of large numbers of Antarctic fur seals.

Marine litter is even found on the floor of the Arctic ocean.

Great Pacific Garbage Patch

Once waterborne, debris becomes mobile. Flotsam can be blown by the wind, or follow the flow of ocean currents, often ending up in the middle of oceanic gyres where currents are weakest. The Great Pacific Garbage Patch is one such example of this, comprising a vast region of the North Pacific Ocean rich with anthropogenic wastes. Estimated to be double the size of Texas, the area contains more than 3 million tons of plastic. In fact, patches may be large enough to be picked up by satellites. For example, when the Malaysian Flight MH370, disappeared in 2014, satellites were scanning the oceans surface for any sign of it, and instead of finding debris from the plane they came across floating garbage. The gyre contains approximately six pounds of plastic for every pound of plankton. The oceans may contain as much as one hundred million tons of plastic. Its estimated that each garbage patch in the ocean have up to one million tons of trash swirling around in them, sometimes extending down to around one hundred feet below the surface. Some items that have been extracted from these garbage patches are: a drum of hazardous chemicals, plastic hangers, tires, cable cords, a ton of tangled netting etc.

Over 40% of oceans are classified as subtropical gyres, a fourth of the planets surface area has become an accumulator of floating plastic debris.

Islands situated within gyres frequently have coastlines flooded by waste that washes ashore; prime examples are Midway and Hawaii. Clean-up teams around the world patrol beaches to attack this environmental threat.

Environmental Impact

Remains of an albatross containing ingested flotsam

Many animals that live on or in the sea consume flotsam by mistake, as it often looks similar to their natural prey. Bulky plastic debris may become permanently lodged in the digestive tracts of these animals, blocking the passage of food and causing death through starvation or infection. Tiny floating plastic particles also resemble zooplankton, which can lead filter feeders to consume them and cause them to enter the ocean food chain. In samples taken from the North Pacific Gyre in 1999 by the Algalita Marine Research Foundation, the mass of plastic exceeded that of zooplankton by a factor of six.

A turtle trapped in a ghost net, an abandoned fishing net

Toxic additives used in plastic manufacturing can leach into their surroundings when exposed to water. Waterborne hydrophobic pollutants collect and magnify on the surface of plastic debris, thus making plastic more deadly in the ocean than it would be on land. Hydrophobic contaminants bioaccumulate in fatty tissues, biomagnifying up the food chain and pressuring apex predators and humans. Some plastic additives disrupt the endocrine system when consumed; others can suppress the immune system or decrease reproductive rates.

The hydrophobic nature of plastic surfaces stimulates rapid formation of biofilms, which support a wide range of metabolic activities, and drive succession of other micro- and macro-organisms.

Concern among experts has grown since the 2000s that some organisms have adapted to live on floating plastic debris, allowing them to disperse with ocean currents and thus potentially become invasive species in distant ecosystems. Research in 2014 in the waters around Australia confirmed a wealth of such colonists, even on tiny flakes, and also found thriving ocean bacteria eating into the plastic to form pits and grooves. These researchers showed that "plastic biodegradation is occurring at the sea surface" through the action of bacteria, and noted that this is congruent with a new body of research on such bacteria. Their finding is also congruent with the other major research undertaken in 2014, which sought to answer the riddle of the overall lack of build up of floating plastic in the oceans, despite ongoing high levels of dumping. Plastics were found as microfibres in core samples drilled from sediments at the bottom of the deep ocean. The cause of such widespread deep sea deposition has yet to be determined.

Not all anthropogenic artifacts placed in the oceans are harmful. Iron and concrete structures typically do little damage to the environment because they generally sink to the bottom and become immobile, and at shallow depths they can even provide scaffolding for artificial reefs. Ships and subway cars have been deliberately sunk for that purpose.

Debris Removal

Skimmer boat used to remove floating debris and trash from the Potomac and Anacostia rivers

Techniques for collecting and removing marine (or riverine) debris include the use of debris skimmer boats *(pictured)*. Devices such as these can be used where floating debris presents a danger to navigation. For example, the US Army Corps of Engineers removes 90 tons of "drifting material" from San Francisco Bay every month. The Corps has been doing this work since 1942, when a seaplane carrying Admiral Chester W. Nimitz collided with a piece of floating debris and sank, costing the life of its pilot. Once debris becomes "beach litter", collection by hand and specialized beach-cleaning machines are used to gather the debris.

Elsewhere, "trash traps" are installed on small rivers to capture waterborne debris before it reaches the sea. For example, South Australia's Adelaide operates a number of such traps, known as "trash racks" or "gross pollutant traps" on the Torrens River, which flows (during the wet season) into Gulf St Vincent.

In lakes or near the coast, manual removal can also be used. Project AWARE for example promotes the idea of letting dive clubs clean up litter, for example as a diving exercise.

On the sea, the removal of artificial debris (i.e. plastics) is still in its infancy. However some projects have been started which used ships with nets (Kaisei and New Horizon) to catch some plastics, primarily for research purposes. Another method to gather artificial litter has been proposed by Boyan Slat. He suggested using platforms with arms to gather the debris, situated inside the current of gyres.

Another issue is that removing marine debris from our oceans can potentially cause more harm than good. Cleaning up micro-plastics could also accidentally take out plankton, which are the main lower level food group for the marine food chain and over half of the photosynthesis on earth. One of the most efficient and cost effective ways to help reduce the amount of plastic entering our oceans is to not participate in using singe use plastics, avoid plastic bottled drinks such as water bottles, use reusable shopping bags, and to buy products with reusable packaging.

Plastic-to-fuel Conversion Strategy

The Clean Oceans Project (TCOP) promotes conversion of the plastic waste into valuable liquid fuels, including gasoline, diesel and kerosene, using plastic-to-fuel conversion technology developed by Blest Co. Ltd., a Japanese environmental engineering company. TCOP plans to educate local communities and create a financial incentive for them to recycle plastic, keep their shorelines clean, and minimize plastic waste.

Laws and Treaties

Ocean dumping is controlled by international law, including:

- The London Convention (1972) – a United Nations agreement to control ocean dumping
- MARPOL 73/78 – a convention designed to minimize pollution of the seas, including dumping, oil and exhaust pollution

European Law

In 1972 and 1974, conventions were held in Oslo and Paris respectively, and resulted in the passing of the OSPAR Convention, an international treaty controlling marine pollution in the north-east Atlantic Ocean. The Barcelona Convention protects the Mediterranean Sea. The Water Framework Directive of 2000 is a European Union directive committing EU member states to free inland

and coastal waters from human influence. In the United Kingdom, the Marine and Coastal Access Act 2009 is designed to "ensure clean healthy, safe, productive and biologically diverse oceans and seas, by putting in place better systems for delivering sustainable development of marine and coastal environment".

A sign above a sewer in Colorado Springs warning people to not pollute the local stream by dumping. Eighty percent of marine debris reaches the sea via rivers.

United States Law

In 1972, the United States Congress passed the Ocean Dumping Act, giving the Environmental Protection Agency power to monitor and regulate the dumping of sewage sludge, industrial waste, radioactive waste and biohazardous materials into the nation's territorial waters. The Act was amended sixteen years later to include medical wastes. It is illegal to dispose of any plastic in US waters.

Ownership

Property law, admiralty law and the law of the sea may be of relevance when lost, mislaid, and abandoned property is found at sea. Salvage law rewards salvors for risking life and property to rescue the property of another from peril. On land the distinction between deliberate and accidental loss led to the concept of a "treasure trove". In the United Kingdom, shipwrecked goods should be reported to a Receiver of Wreck, and if identifiable, they should be returned to their rightful owner.

Activism

A large number of groups and individuals are active in preventing or educating about marine debris. For example, 5 Gyres is an organization aimed at reducing plastics pollution in the oceans, and was one of two organizations that recently researched the Great Pacific Garbage Patch. Heal the Bay is another organization, focusing on protecting California's Santa Monica Bay, by sponsoring beach cleanup programs along with other activities. Marina DeBris is an artist focusing most of her recent work on educating people about beach trash. Interactive sites like Adrift demonstrate where marine plastic is carried, over time, on the worlds ocean currents. On April 11, 2013 in order to create awareness, artist Maria Cristina Finucci founded The Garbage patch state at UNESCO –Paris in front of Director General Irina Bokova . First of a series of events under the patronage of UNESCO and of Italian Ministry of the Environment.

Forty-eight plastics manufacturers from 25 countries, are members of the Global Plastic Associations for solutions on Marine Litter, have made the pledge to help prevent marine debris and to encourage recycling.

Chemical Waste

Chemical Waste Bin (Chemobox)

Chemical waste is a waste that is made from harmful chemicals (mostly produced by large factories). Chemical waste may fall under regulations such as COSHH in the United Kingdom, or the Clean Water Act and Resource Conservation and Recovery Act in the United States. In the U.S., the Environmental Protection Agency (EPA) and the Occupational Safety and Health Administration (OSHA), as well as state and local regulations also regulate chemical use and disposal. Chemical waste may or may not be classed as hazardous waste. A chemical hazardous waste is a solid, liquid, or gaseous material that displays either a "Hazardous Characteristic" or is specifically "listed" by name as a hazardous waste. There are four characteristics chemical wastes may have to be considered as hazardous. These are Ignitability, Corrosivity, Reactivity, and Toxicity. This type of hazardous waste must be categorized as to its identity, constituents, and hazards so that it may be safely handled and managed. Chemical waste is a broad term and encompasses many types of materials. Consult the Material Safety Data Sheet (MSDS), Product Data Sheet or Label for a list of constituents. These sources should state weather this chemical waste is a waste that needs special disposal.

Guidance for Disposal of Laboratory Chemical Wastes

In the laboratory, chemical wastes are usually segregated on-site into appropriate waste carboys, and disposed by a specialist contractor in order to meet safety, health, and legislative requirements.

Innocuous aqueous waste (such as solutions of sodium chloride) may be poured down the sink. Some chemicals are washed down with excess water. This includes: concentrated and dilute acids and alkalis, harmless soluble inorganic salts (all drying agents), alcohols containing salts, hypochlorite solutions, fine (tlc grade) silica and alumina. Aqueous waste containing toxic compounds are collected separately

Chemical Waste Disposal Guideline

Innocuous aqueous waste	Organic Solvent	Red List	Solid Waste
• **Acid** (pH<4) • **Alkali** (pH>10) • Harmless soluble inorganic salt • Alcohol containing salt • Hypochlorite solution • Fine (tlc grade) silica and alumina These chemicals should be washed down with excess water.	• **Chlorinated** Example: DCM, Chloroform, Chlorobenzene etc. • **Non-Chlorinated** Example: THF, ethyl acetate, hexane, toluene, methanol, etc.	• Compounds with transitional metals • Biocides • Cyanides • Mineral oils and hydrocarbons • Poisonous organosilicon compounds • Metal phosphides • Phosphorus element • Fluorides and nitrites.	• Lightly contaminated Example: Gloves, empty vials/centrifuge . **Broken Glassware** Broken glassware are usually collected in plastic-lined cardboard boxes for landfilling. Due to contamination, they are usually not suitable for recycling.

Chemical waste category that should be followed for proper packaging, labelling, and disposal of chemical waste.

Waste elemental mercury, spent acids and bases may be collected separately for recycling.

Waste organic solvents are separated into chlorinated and non-chlorinated solvent waste. Chlorinated solvent waste is usually incinerated at high temperature to minimize the formation of dioxins. Non-chlorinated solvent waste can be burned for energy recovery.

In contrast to this, chemical materials on the "Red List" should never be washed down a drain. This list includes: compounds with transitional metals, biocides, cyanides, mineral oils and hydrocarbons, poisonous organosilicon compounds, metal phosphides, phosphorus element, and fluorides and nitrites.

Moreover, the Environmental Protection Agency (EPA) prohibits disposing certain materials down any UVM drain. Including flammable liquids, liquids capable of causing damage to wastewater facilities (this can be determined by the pH), highly viscous materials capable of causing an obstruction in the wastewater system, radioactive materials, materials that have or create a strong odor, wastewater capable of significantly raising the temperature of the system, and pharmaceuticals or endocrine disruptors.

Broken glassware are usually collected in plastic-lined cardboard boxes for landfilling. Due to contamination, they are usually not suitable for recycling. Similarly, used hypodermic needles are collected as sharps and are incinerated as medical waste.

Chemical Compatibility Guideline

Many chemicals may react adversely when combined. It's recommended that incompatible chemicals are stored in separate areas of the lab.

Acids should be separated from alkalis, metals, cyanides, sulfides, azides, phosphides, and oxidizers. The reason being, when combined acids with these type of compounds, violent exothermic reaction can occur possibly causing flammable gas, and in some cases explosions.

Oxidizers should be separated from acids, organic materials, metals, reducing agents, and ammonia. This is because when combined oxidizers with these type of compounds, inflammable, and sometimes toxic compounds can occur.

Container Compatibility

When disposing hazardous laboratory chemical waste, chemical compatibility must be considered. For safe disposal, the container must be chemically compatible with the material it will hold. Chemicals must not react with, weaken, or dissolve the container or lid. Acids or bases should not be stored in metal. Hydrofluoric acid should not store in glass. Gasoline (solvents) should not store or transport in lightweight polyethylene containers such as milk jugs. Moreover, the Chemical Compatibility Guidelines should be considered for more detailed information.

Laboratory Waste Containers

Packaging, labelling, storage are the three requirements for disposing chemical waste.

Packaging

How to properly label, package, and store chemical waste safely.

For packaging, chemical liquid waste containers should only be filled up to 75% capacity to allow for vapour expansion and to reduce potential spills which could occur from moving overfilled containers. Container material must be compatible with the stored hazardous waste. Finally, wastes must not be packaged in containers that improperly identify other nonexisting hazards.

In addition to the general packaging requirements mentioned above, incompatible materials should never be mixed together in a single container. Wastes must be stored in containers compatible with the chemicals stored as mentioned in the container compatibility section. Solvent safety cans should to be used to collect and temporarily store large volumes (10-20 litres) of flammable organic waste solvents, precipitates, solids or other non-fluid wastes should not be mixed into safety cans.

Labelling

Label all containers with the group name from the chemical waste category and an itemized list of the contents. All chemicals or anything contaminated with chemicals posing a significant hazard. All waste must be appropriately packaged.

Storage

When storing chemical wastes, the containers must be in good condition and should remain closed unless waste is being added. Hazardous waste must be stored safely prior to removal from the laboratory and should not be allowed to accumulate. Container should be sturdy and leakproof, also has to be labeled. All liquid waste must be stored in leakproof containers with a screw- top or other secure lid. Snap caps, mis-sized caps, parafilm and other loose fitting lids are not acceptable. If necessary, transfer waste material to a container that can be securely closed. Keep waste containers closed except when adding waste. Secondary containment should be in place to capture spills and leaks from the primary container, segregate incompatible hazardous wastes, such as acids and bases.

Mapping of Chemical Waste in the United States

TOXMAP is a Geographic Information System (GIS) from the Division of Specialized Information Services of the United States National Library of Medicine (NLM) that uses maps of the United States to help users visually explore data from the United States Environmental Protection Agency's (EPA) Toxics Release Inventory and Superfund Basic Research Programs. TOXMAP is a resource funded by the US Federal Government. TOXMAP's chemical and environmental health information is taken from NLM's Toxicology Data Network (TOXNET) and PubMed, and from other authoritative sources.

Chemical Waste in Canadian Aquaculture

Green Sea Urchin or S. droebacheinsis

Chemical waste in our oceans is becoming a major issue for the marine life. There have been many studies conducted to try an prove the effects of these chemical in our oceans. In Canada, many of the studies concentrated on the Atlantic provinces, where fishing and aquaculture are an important part of the economy. In New Brunswick, a study was done on the sea urchin in an attempt to identify the effects of toxic and chemical waste on life beneath the ocean, specifically the wasted from the salmon farms. Sea urchins were used to check the levels of metals in the environment.

It is advantageous to use green sea urchins, Strongylocentrotus droebachiensis, because they are widely distributed, abundant in many locations, and easily accessible. By investigating the concentrations of metals in the green sea urchins, the impacts of produced chemicals from salmon aquaculture activity could be assessed and detected. Samples were taken at 25m intervals along a transect in the direction of the main tidal flow. The study found that there was impacts to at least 75m based on the intestine metal concentrations. So based on this study it is clear that the metals are contaminating the oceans and negatively affecting aquatic life.

Uranium in Ground and Surface Water in Canada

Another issue regarding chemical waste is the potential risk of surface and groundwater contamination by the heavy metals and radionuclides leached from uranium waste-rock piles (UWRP) A Radionuclide is an atom that has excess nuclear energy, making it unstable. Uranium waste-rock piles refers to Uranium mining, which is the process of extraction of uranium ore from the ground. . An example of such threats is in Saskatchewan, Uranium mining and ore processing (milling) can pose a threat to the environment. In open pit mining, large amounts of materials are excavated and disposed off in waste-rock piles. Waste-rock piles from the Uranium mining industry can contain several heavy metals and contaminants that may become mobile under certain conditions. Environmental contaminants may include acid mine drainage, higher concentrations of radionuclides, and non-radioactive metals/metalloids (i.e. As, Mo, Ni, Cu, Zn).

The leachability of heavy metals and radionuclide from UWRP plays a significant role in determining their potential environmental risks to surrounding surface and groundwater. Substantial differences in the solid-phase partitioning and chemical leachability of Ni and U were observed in the investigated UWRP lithological materials and background organic-rich lake sediment. For Instance, in the uranium-mining district of Northern Saskatchewan, Canada, the sequential extraction results showed that a significant amount of Ni (Nickel) was present in the non-labile residual fraction, while Uranium was mostly distributed in the moderately labile fractions. Although Nickel was much less labile than Uranium, the observed Nickel exceeded Uranium concentrations in leaching]].The observed Nickel and Uranium concentrations were relatively high in the underlying organic-rich lake sediment. Expressed as the percentage of total metal content, potential leachability decreased in the order U > Ni. Data suggest that these elements could potentially migrate to the water table below the UWRP. Detailed information regarding the solid-phase distribution of contaminants in the UWRP is critical to understand the potential for their environmental transport and mobility.

The most visible civilian use of uranium is as the thermal power source used in nuclear power plants

Acid Mine Drainage

Acid mine drainage, acid and metalliferous drainage (AMD), or acid rock drainage (ARD) refers to the outflow of acidic water from metal mines or coal mines.

Yellow boy in a stream receiving acid drainage from surface coal mining.

Rocks stained by acid mine drainage on Shamokin Creek

Acid rock drainage occurs naturally within some environments as part of the rock weathering process but is exacerbated by large-scale earth disturbances characteristic of mining and other large construction activities, usually within rocks containing an abundance of sulfide minerals. Areas where the earth has been disturbed (e.g. construction sites, subdivisions, and transportation corridors) may create acid rock drainage. In many localities, the liquid that drains from coal stocks, coal handling facilities, coal washeries, and coal waste tips can be highly acidic, and in such cases it is treated as acid rock drainage.

The same type of chemical reactions and processes may occur through the disturbance of acid sulfate soils formed under coastal or estuarine conditions after the last major sea level rise, and constitutes a similar environmental hazard.

Nomenclature

Historically, the acidic discharges from active or abandoned mines were called acid mine drainage, or AMD. The term acid rock drainage, or ARD, was introduced in the 1980s and 1990s to indicate that acidic drainage can originate from sources other than mines. For example, a paper presented in 1991 at a major international conference on this subject was titled: "The Prediction of Acid Rock Drainage - Lessons from the Database" Both AMD and ARD refer to low pH or acidic waters caused by the oxidation of sulfide minerals, though ARD is the more generic name.

In cases where drainage from a mine is not acidic and has dissolved metals or metalloids, or was originally acidic, but has been neutralized along its flow path, then it is described as "Neutral Mine Drainage", "Mining-Influenced Water" or otherwise. None of these other names have gained general acceptance.

Occurrence

In this case, the pyrite has dissolved away yielding a cube shape and residual gold. This break down is the main driver of acid mine drainage.

Sub-surface mining often progresses below the water table, so water must be constantly pumped out of the mine in order to prevent flooding. When a mine is abandoned, the pumping ceases, and water floods the mine. This introduction of water is the initial step in most acid rock drainage situations. Tailings piles or ponds, mine waste rock dumps, and coal spoils are also an important source of acid mine drainage.

After being exposed to air and water, oxidation of metal sulfides (often pyrite, which is iron-sulfide) within the surrounding rock and overburden generates acidity. Colonies of bacteria and archaea greatly accelerate the decomposition of metal ions, although the reactions also occur in an abiotic environment. These microbes, called extremophiles for their ability to survive in harsh conditions, occur naturally in the rock, but limited water and oxygen supplies usually keep their numbers low. Special extremophiles known as Acidophiles especially favor the low pH levels of abandoned mines. In particular, Acidithiobacillus *ferrooxidans* is a key contributor to pyrite oxidation.

Metal mines may generate highly acidic discharges where the ore is a sulfide mineral or is associated with pyrite. In these cases the predominant metal ion may not be iron but rather zinc, copper, or nickel. The most commonly mined ore of copper, chalcopyrite, is itself a copper-iron-sulfide and occurs with a range of other sulfides. Thus, copper mines are often major culprits of acid mine drainage.

At some mines, acidic drainage is detected within 2–5 years after mining begins, whereas at other mines, it is not detected for several decades. In addition, acidic drainage may be generated for decades or centuries after it is first detected. For this reason, acid mine drainage is considered a serious long-term environmental problem associated with mining.

Chemistry

The chemistry of oxidation of pyrites, the production of ferrous ions and subsequently ferric ions, is very complex, and this complexity has considerably inhibited the design of effective treatment options.

Although a host of chemical processes contribute to acid mine drainage, pyrite oxidation is by far the greatest contributor. A general equation for this process is:

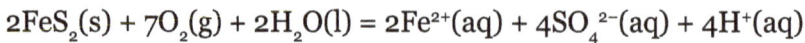

$$2FeS_2(s) + 7O_2(g) + 2H_2O(l) = 2Fe^{2+}(aq) + 4SO_4^{2-}(aq) + 4H^+(aq)$$

The oxidation of the sulfide to sulfate solubilizes the ferrous iron (iron(II)), which is subsequently oxidized to ferric iron (iron(III)):

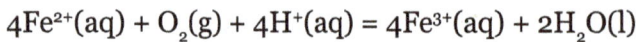

$$4Fe^{2+}(aq) + O_2(g) + 4H^+(aq) = 4Fe^{3+}(aq) + 2H_2O(l)$$

Either of these reactions can occur spontaneously or can be catalyzed by microorganisms that derive energy from the oxidation reaction. The ferric cations produced can also oxidize additional pyrite and reduce into ferrous ions:

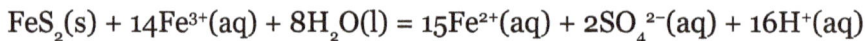

$$FeS_2(s) + 14Fe^{3+}(aq) + 8H_2O(l) = 15Fe^{2+}(aq) + 2SO_4^{2-}(aq) + 16H^+(aq)$$

The net effect of these reactions is to release H^+, which lowers the pH and maintains the solubility of the ferric ion.

Effects

Effects on pH

Water temperatures as high as 47 °C have been measured underground at the Iron Mountain Mine, and the pH can be as low as -3.6.

Rio Tinto in Spain.

Organisms which cause acid mine drainage can thrive in waters with pH very close to zero. Negative pH occurs when water evaporates from already acidic pools thereby increasing the concentration of hydrogen ions.

About half of the coal mine discharges in Pennsylvania have pH under 5. However, a significant portion of mine drainage in both the bituminous and anthracite regions of Pennsylvania is alkaline, because limestone in the overburden neutralizes acid before the drainage emanates.

Acid rock drainage has recently been a hindrance to the completion of the construction of Interstate 99 near State College, Pennsylvania. However, this acid rock drainage didn't come from a mine; rather, it was produced by oxidation of pyrite-rich rock which was unearthed during a road cut and then used as filler material in the I-99 construction. A similar situation developed at the Halifax airport in Canada. It is from these and similar experiences that the term acid *rock* drainage has emerged as being preferable to acid *mine* drainage, thereby emphasizing the general nature of the problem.

Yellow Boy

When the pH of acid mine drainage is raised past 3, either through contact with fresh water or neutralizing minerals, previously soluble iron(III) ions precipitate as iron(III) hydroxide, a yellow-orange solid colloquially known as *yellow boy*. Other types of iron precipitates are possible, including iron oxides and oxyhydroxides. All these precipitates can discolor water and smother plant and animal life on the streambed, disrupting stream ecosystems (a specific offense under the Fisheries Act in Canada). The process also produces additional hydrogen ions, which can further decrease pH. In some cases, the concentrations of iron hydroxides in yellow boy are so high, the precipitate can be recovered for commercial use in pigments.

Trace Metal and Semi-Metal Contamination

Many acid rock discharges also contain elevated levels of potentially toxic metals, especially nickel and copper with lower levels of a range of trace and semi-metal ions such as lead, arsenic, aluminium, and manganese. The elevated levels of heavy metals can only be dissolved in waters that have a low pH, as is found in the acidic waters produced by pyrite oxidation. In the coal belt around the south Wales valleys in the UK highly acidic nickel-rich discharges from coal stocking sites have proved to be particularly troublesome.

Identification and Prediction

In a mining setting it is leading practice to carry out a geochemical assessment of mine materials during the early stages of a project to determine the potential for AMD. The geochemical assessment aims to map the distribution and variability of key geochemical parameters, acid generating and element leaching characteristics.

The assessment may include:

1. Sampling;

2. Static geochemical testwork (e.g. acid-base accounting, sulfur speciation);

3. Kinetic geochemical testwork - Conducting oxygen consumption tests, such as the OxCon, to quantify acidity generation rates

4. Modelling of oxidation, pollutant generation and release; and

5. Modelling of material composition.

Treatment

Oversight

In the United Kingdom, many discharges from abandoned mines are exempt from regulatory control. In such cases the Environment Agency working with partners such as the Coal Authority have provided some innovative solutions, including constructed wetland solutions such as on the River Pelenna in the valley of the River Afan near Port Talbot and the constructed wetland next to the River Neath at Ynysarwed.

Although abandoned underground mines produce most of the acid mine drainage, some recently mined and reclaimed surface mines have produced ARD and have degraded local ground-water and surface-water resources. Acidic water produced at active mines must be neutralized to achieve pH 6-9 before discharge from a mine site to a stream is permitted.

In Canada, work to reduce the effects of acid mine drainage is concentrated under the Mine Environment Neutral Drainage (MEND) program. Total liability from acid rock drainage is estimated to be between \$2 billion and \$5 billion CAD. Over a period of eight years, MEND claims to have reduced ARD liability by up to \$400 million CAD, from an investment of \$17.5 million CAD.

Methods

Lime Neutralization

By far, the most commonly used commercial process for treating acid mine drainage is lime precipitation in a high-density sludge (HDS) process. In this application, a slurry of lime is dispersed into a tank containing acid mine drainage and recycled sludge to increase water pH to about 9. At this pH, most toxic metals become insoluble and precipitate, aided by the presence of recycled sludge. Optionally, air may be introduced in this tank to oxidize iron and manganese and assist in their precipitation. The resulting slurry is directed to a sludge-settling vessel, such as a clarifier. In that vessel, clean water will overflow for release, whereas settled metal precipitates (sludge) will be recycled to the acid mine drainage treatment tank, with a sludge-wasting side stream. A number of variations of this process exist, as dictated by the chemistry of ARD, its volume, and other factors. Generally, the products of the HDS process also contain gypsum and unreacted lime, which enhance both its settleability and resistance to re-acidification and metal mobilization.

Less complex variants of this process, such as simple lime neutralization, may involve no more than a lime silo, mixing tank and settling pond. These systems are far less costly to build, but are also less efficient (i.e., longer reaction times are required, and they produce a discharge with higher trace metal concentrations, if present). They would be suitable for relatively small flows or less complex acid mine drainage.

Calcium Silicate Neutralization

A calcium silicate feedstock, made from processed steel slag, can also be used to neutralize active acidity in AMD systems by removing free hydrogen ions from the bulk solution, thereby increasing pH. As the silicate anion captures H^+ ions (raising the pH), it forms monosilicic acid (H_4SiO_4), a neutral solute. Monosilicic acid remains in the bulk solution to play many roles in correcting the adverse effects of acidic conditions. In the bulk solution, the silicate anion is very active in neutralizing H^+ cations in the soil solution. While its mode-of-action is quite different from limestone, the ability of calcium silicate to neutralize acid solutions is equivalent to limestone as evidenced by its CCE value of 90-100% and its relative neutralizing value of 98%.

In the presence of heavy metals, calcium silicate reacts in a different manner than limestone. As limestone raises the pH of the bulk solution, and if heavy metals are present, precipitation of the metal hydroxides (with extremely low solubilities) is normally accelerated and the potential of armoring of limestone particles increases significantly. In the calcium silicate aggregate, as silicic acid species are absorbed onto the metal surface, the development of silica layers (mono- and bi-layers) lead to the formation of colloidal complexes with neutral or negative surface charges. These negatively charged colloids create an electrostatic repulsion with each other (as well as with the negatively charged calcium silicate granules) and the sequestered metal colloids are stabilized and remain in a dispersed state - effectively interrupting metal precipitation and reducing vulnerability of the material to armoring.

Carbonate Neutralization

Generally, limestone or other calcareous strata that could neutralize acid are lacking or deficient at sites that produce acidic rock drainage. Limestone chips may be introduced into sites to create a neutralizing effect. Where limestone has been used, such as at Cwm Rheidol in mid Wales, the positive impact has been much less than anticipated because of the creation of an insoluble calcium sulfate layer on the limestone chips, binding the material and preventing further neutralization.

Ion Exchange

Cation exchange processes have previously been investigated as a potential treatment for acid mine drainage. The principle is that an ion exchange resin can remove potentially toxic metals (cationic resins), or chlorides, sulfates and uranyl sulfate complexes (anionic resins) from mine water. Once the contaminants are adsorbed, the exchange sites on resins must be regenerated, which typically requires acidic and basic reagents and generates a brine that contains the pollutants in a concentrated form. A South African company that won the 2013 IChemE (ww.icheme. org) award for water management and supply (treating AMD) have developed a patented ion-exchange process that treats mine effluents (and AMD) economically.

Constructed Wetlands

Constructed wetlands systems have been proposed during the 1980s to treat acid mine drainage generated by the abandoned coal mines in Eastern Appalachia. Generally, the wetlands receive near-neutral water, after it has been neutralized by (typically) a limestone-based treatment process. Metal precipitation occurs from their oxidation at near-neutral pH, complexation with organic matter, precipitation as carbonates or sulfides. The latter results from sediment-borne anaerobic bacteria capable of reverting sulfate ions into sulfide ions. These sulfide ions can then bind with heavy metal ions, precipitating heavy metals out of solution and effectively reversing the entire process.

The attractiveness of a constructed wetlands solution lies in its relative low cost. They are limited by the metal loads they can deal with (either from high flows or metal concentrations), though current practitioners have succeeded in developing constructed wetlands that treat high volumes and/or highly acidic water (with adequate pre-treatment). Typically, the effluent from constructed wetland receiving near-neutral water will be well-buffered at between 6.5-7.0 and can readily be discharged. Some of metal precipitates retained in sediments are unstable when exposed to oxygen (e.g., copper sulfide or elemental selenium), and it is very important that the wetland sediments remain largely or permanently submerged.

An example of an effective constructed wetland is on the Afon Pelena in the River Afan valley above Port Talbot where highly ferruginous discharges from the Whitworth mine have been successfully treated.

Precipitation of Metal Sulfides

Most base metals in acidic solution precipitate in contact with free sulfide, e.g. from H_2S or NaHS. Solid-liquid separation after reaction would produce a base metal-free effluent that can be discharged or further treated to reduce sulfate, and a metal sulfide concentrate with possible economic value.

As an alternative, several researchers have investigated the precipitation of metals using biogenic sulfide. In this process, Sulfate-reducing bacteria oxidize organic matter using sulfate, instead of oxygen. Their metabolic products include bicarbonate, which can neutralize water acidity, and hydrogen sulfide, which forms highly insoluble precipitates with many toxic metals. Although promising, this process has been slow in being adopted for a variety of technical reasons.

Technologies

Many technologies exist for the treatment of AMD from traditional high cost water treatment plants to simple in situ water treatment reagent dosing methods.

Metagenomic Study of Acid Mine Drainage

With the advance of Large-scale sequencing strategies, genomes of microorganisms in the acid mine drainage community are directly sequenced from the environment. The nearly full genomic constructs allows new understanding of the community and able to reconstruct their metabolic pathways. Our knowledge of Acidophiles in acid mine drainage remains rudimentary: we know of many more species associated with ARD than we can establish roles and functions.

Microbes and Drug Discovery

Scientists have recently begun to explore acid mine drainage and mine reclamation sites for unique soil bacteria capable of producing new pharmaceutical leads. Soil microbes have long been a source for effective drugs and new research, such as that conducted at the Center for Pharmaceutical Research and Innovation, suggests these extreme environments to be an untapped source for new discovery.

List of Selected Acid Mine Drainage Sites Worldwide

This list includes both mines producing acid mine drainage and river systems significantly affected by such drainage. It is by no means complete, as worldwide, several thousands of such sites exist.

References

- Parker, Laura. "With Millions of Tons of Plastic in Oceans, More Scientists Studying Impact." National Geographic. National Geographic Society, 13 June 2014. Web. 03 Apr. 2016.

- Hallam, Bill (April–May 2010). "Techniques for Efficient Hazardous Chemicals Handling and Disposal". Pollution Equipment News. p. 13. Retrieved 10 March 2016.

- "LABORATORY CHEMICAL WASTE MANAGEMENT GUIDELINES" (PDF). Environmental Health and Radiation Safety University of Pennsylvania. Retrieved 10 March 2016.

- "Waste - Disposal of Laboratory Wastes (GUIDANCE) | Current Staff | University of St Andrews". www.st-andrews.ac.uk. Retrieved 2016-02-04.

- Macrae, Fiona (12 February 2015). "Eight million tons of plastic is dumped at sea each year.". Daily Mail. Retrieved 21 February 2015.

- Edgar B. Herwick III (29 July 2015). "Explosive Beach Objects-- Just Another Example Of Massachusetts' Charm". WGBH news. PBS. Retrieved 4 August 2015.

- "Millimeter-Sized Marine Plastics: A New Pelagic Habitat for Microorganisms and Invertebrates". PLoS ONE. 18 June 2014. Retrieved 2015-01-26.

- "Overview of acid mine drainage impacts in the West Rand Goldfield". Presentation to DG of DWAF. 2 February 2009. Archived from the original on 2012-03-13. Retrieved 2 July 2014.

- David Falchek (26 December 2012). "Old Forge borehole drains mines for 50 years". The Scranton Times Tribune. Retrieved 18 March 2013.

- "MBARI News Release: MBARI research shows where trash accumulates in the deep sea". Monterey Bay Aquarium Research Institute. 2013. Retrieved 2013-07-07.

- Laboratory, National Research Council (US) Committee on Prudent Practices in the (2011-01-01). "Management of Waste". Retrieved 10 March 2016.

- Hammarstrom, Jane M.; Philip L. Sibrell; Harvey E. Belkin. "Characterization of limestone reacted with acid-mine drainage" (PDF). Applied Geochemistry (18): 1710–1714. Retrieved 30 March 2011.

- André Sobolewski. "Constructed wetlands for treatment of mine drainage - Coal-generated AMD". Wetlands for the Treatment of Mine Drainage. Retrieved 2010-12-12.

4

Water Quality Parameters Model

Water quality modeling commits to the prediction of water pollution using mathematical simulation techniques. A typical water quality model is a collection of formulations representing physical mechanisms that determine the position of pollutants in a water body. The chapter strategically encompasses and incorporates the major components of water quality modeling.

Hydrological Transport Model

An hydrological transport model is a mathematical model used to simulate river or stream flow and calculate water quality parameters. These models generally came into use in the 1960s and 1970s when demand for numerical forecasting of water quality was driven by environmental legislation, and at a similar time widespread access to significant computer power became available. Much of the original model development took place in the United States and United Kingdom, but today these models are refined and used worldwide.

River in Madagascar relatively free of sediment load

There are dozens of different transport models that can be generally grouped by pollutants addressed, complexity of pollutant sources, whether the model is steady state or dynamic, and time period modeled. Another important designation is whether the model is distributed (i.e. capable of predicting multiple points within a river) or lumped. In a basic model, for example, only one pollutant might be addressed from a simple point discharge into the receiving waters. In the most complex of models, various line source inputs from surface runoff might be added to multiple point sources, treating a variety of chemicals plus sediment in a dynamic environment including vertical river stratification and interactions of pollutants with in-stream biota. In addition watershed groundwater may also be included. The model is termed "physically based" if its parameters can be measured in the field.

Often models have separate modules to address individual steps in the simulation process. The most common module is a subroutine for calculation of surface runoff, allowing variation in land use type, topography, soil type, vegetative cover, precipitation and land management practice (such as the application rate of a fertilizer). The concept of hydrological modeling can be extended to other environments such as the oceans, but most commonly (and in this article) the subject of a river watershed is generally implied.

History

In 1850, T. J. Mulvany was probably the first investigator to use mathematical modeling in a stream hydrology context, although there was no chemistry involved. By 1892 M.E. Imbeau had conceived an event model to relate runoff to peak rainfall, again still with no chemistry. Robert E. Horton's seminal work on surface runoff along with his coupling of quantitative treatment of erosion laid the groundwork for modern chemical transport hydrology.

Types of Hydrological Transport Models

Physically Based Models

Physically based models (sometimes known as deterministic, comprehensive or process-based models) try to represent the physical processes observed in the real world. Typically, such models contain representations of surface runoff, subsurface flow, evapotranspiration, and channel flow, but they can be far more complicated. "Large scale simulation experiments were begun by the U.S. Army Corps of Engineers in 1953 for reservoir management on the main stem of the Missouri River". This, and other early work that dealt with the River Nile and the Columbia River are discussed, in a wider context, in a book published by the Harvard Water Resources Seminar, that contains the sentence just quoted. Another early model that integrated many submodels for basin chemical hydrology was the Stanford Watershed Model (SWM). The SWMM (Storm Water Management Model), the HSPF (Hydrological Simulation Program - FORTRAN) and other modern American derivatives are successors to this early work.

In Europe a favoured comprehensive model is the Système Hydrologique Européen (SHE), which has been succeeded by MIKE SHE and SHETRAN. MIKE SHE is a watershed-scale physically based, spatially distributed model for water flow and sediment transport. Flow and transport processes are represented by either finite difference representations of partial differential equations or by derived empirical equations. The following principal submodels are involved:

- Evapotranspiration: Penman-Monteith formalism

- Erosion: Detachment equations for raindrop and overland flow

- Overland and Channel Flow: Saint-Venant equations of continuity and momentum

- Overland Flow Sediment Transport: 2D total sediment load conservation equation

- Unsaturated Flow: Richards equation

- Saturated Flow: Darcy's law and the mass conservation of 2D laminar flow

- Channel Sediment Transport 1D mass conservation equation.

This model can analyze effects of land use and climate changes upon in-stream water quality, with consideration of groundwater interactions.

Worldwide a number of basin models have been developed, among them RORB (Australia), Xinan-jiang (China), Tank model (Japan), ARNO (Italy), TOPMODEL (Europe), UBC (Canada) and HBV (Scandinavia), MOHID Land (Portugal). However, not all these models have a chemistry component. Generally speaking, SWM, SHE and TOPMODEL have the most comprehensive stream chemistry treatment and have evolved to accommodate the latest data sources including remote sensing and geographic information system data.

In the United States, the Corps of Engineers, Engineer Research and Development Center in conjunction with a researchers at a number of universities have developed the Gridded Surface/Subsurface Hydrologic Analysis GSSHA model. GSSHA is widely used in the U.S. for research and analysis by U.S. Army Corps of Engineers districts and larger consulting companies to compute flow, water levels, distributed erosion, and sediment delivery in complex engineering designs. A distributed nutrient and contaminant fate and transport component is undergoing testing. GSSHA input/output processing and interface with GIS is facilitated by the Watershed Modeling System (WMS).

Another model used in the United States and worldwide is V*flo*, a physics-based distributed hydrologic model developed by Vieux & Associates, Inc. V*flo* employs radar rainfall and GIS data to compute spatially distributed overland flow and channel flow. Evapotranspiration, inundation, infiltration, and snowmelt modeling capabilities are included. Applications include civil infrastructure operations and maintenance, stormwater prediction and emergency management, soil moisture monitoring, land use planning, water quality monitoring, and others.

Stochastic Models

These models based on data are black box systems, using mathematical and statistical concepts to link a certain input (for instance rainfall) to the model output (for instance runoff). Commonly used techniques are regression, transfer functions, neural networks and system identification. These models are known as stochastic hydrology models. Data based models have been used within hydrology to simulate the rainfall-runoff relationship, represent the impacts of antecedent moisture and perform real-time control on systems.

Model Components

Surface Runoff Modelling

A key component of a hydrological transport model is the surface runoff element, which allows assessment of sediment, fertilizer, pesticide and other chemical contaminants. Building on the work of Horton, the unit hydrograph theory was developed by Dooge in 1959. It required the presence of the National Environmental Policy Act and kindred other national legislation to provide the impetus to integrate water chemistry to hydrology model protocols. In the early 1970s the U.S.

Environmental Protection Agency (EPA) began sponsoring a series of water quality models in response to the Clean Water Act. An example of these efforts was developed at the Southeast Water Laboratory, one of the first attempts to calibrate a surface runoff model with field data for a variety of chemical contaminants.

Columbia River, which has surface runoff from agriculture and logging

The attention given to surface runoff contaminant models has not matched the emphasis on pure hydrology models, in spite of their role in the generation of stream loading contaminant data. In the United States the EPA has had difficulty interpreting diverse proprietary contaminant models and has to develop its own models more often than conventional resource agencies, who, focused on flood forecasting, have had more of a centroid of common basin models.

Example Applications

Liden applied the HBV model to estimate the riverine transport of three different substances, nitrogen, phosphorus and suspended sediment in four different countries: Sweden, Estonia, Bolivia and Zimbabwe. The relation between internal hydrological model variables and nutrient transport was assessed. A model for nitrogen sources was developed and analysed in comparison with a statistical method. A model for suspended sediment transport in tropical and semi-arid regions was developed and tested. It was shown that riverine total nitrogen could be well simulated in the Nordic climate and riverine suspended sediment load could be estimated fairly well in tropical and semi-arid climates. The HBV model for material transport generally estimated material transport loads well. The main conclusion of the study was that the HBV model can be used to predict material transport on the scale of the drainage basin during stationary conditions, but cannot be easily generalised to areas not specifically calibrated. In a different work, Castanedo et al. applied an evolutionary algorithm to automated watershed model calibration.

The United States EPA developed the DSSAM Model to analyze water quality impacts from land use and wastewater management decisions in the Truckee River basin, an area which include the cities of Reno and Sparks, Nevada as well as the Lake Tahoe basin. The model satisfactorily predicted nutrient, sediment and dissolved oxygen parameters in the river. It is based on a pollutant loading metric called "Total Daily Maximum Load" (TDML). The success of this model contributed to the EPA's commitment to the use of the underlying TDML protocol in EPA's national policy for management of many river systems in the United States.

The DSSAM Model is constructed to allow dynamic decay of most pollutants; for example, total nitrogen and phosphorus are allowed to be consumed by benthic algae in each time step, and the algal communities are given a separate population dynamic in each river reach (e.g. based upon river temperature). Regarding stormwater runoff in Washoe County, the specific elements within a new xeriscape ordinance were analyzed for efficacy using the model. For the varied agricultural uses in the watershed, the model was run to understand the principal sources of impact, and management practices were developed to reduce in-river pollution. Use of the model has specifically been conducted to analyze survival of two endangered species found in the Truckee River and Pyramid Lake: the Cui-ui sucker fish (endangered 1967) and the Lahontan cutthroat trout (threatened 1970).

Groundwater Model

Groundwater models are computer models of groundwater flow systems, and are used by hydrogeologists. Groundwater models are used to simulate and predict aquifer conditions.

Characteristics

An unambiguous definition of "groundwater model" is difficult to give, but there are many common characteristics.

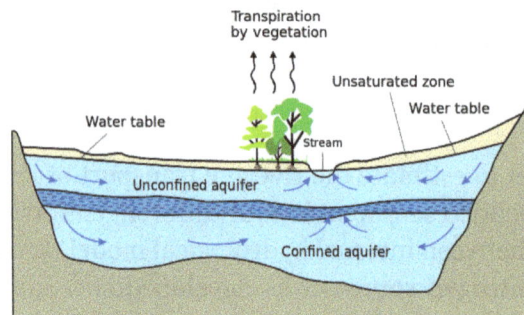

Typical aquifer cross-section

A groundwater model may be a scale model or an electric model of a groundwater situation or aquifer. Groundwater models are used to represent the natural groundwater flow in the environment. Some groundwater models include (chemical) quality aspects of the groundwater. Such groundwater models try to predict the fate and movement of the chemical in natural, urban or hypothetical scenario.

Groundwater models may be used to predict the effects of hydrological changes (like groundwater abstraction or irrigation developments) on the behavior of the aquifer and are often named groundwater simulation models. Also nowadays the groundwater models are used in various water management plans for urban areas.

As the computations in mathematical groundwater models are based on groundwater flow equations, which are differential equations that can often be solved only by approximate methods using a numerical analysis, these models are also called *mathematical, numerical, or computational groundwater models.*

The mathematical or the numerical models are usually based on the real physics the groundwater flow follows. These mathematical equations are solved using numerical codes such as MODFLOW, ParFlow, HydroGeoSphere, OpenGeoSys etc. Various types of *numerical solutions* like the finite difference method and the finite element method are discussed in the article on "Hydrogeology".

Inputs

For the calculations one needs inputs like:

- hydrological inputs,

- operational inputs,

- external conditions: initial and boundary conditions,

- (hydraulic) parameters.

The model may have chemical components like water salinity, soil salinity and other quality indicators of water and soil, for which inputs may also be needed.

Hydrological Inputs

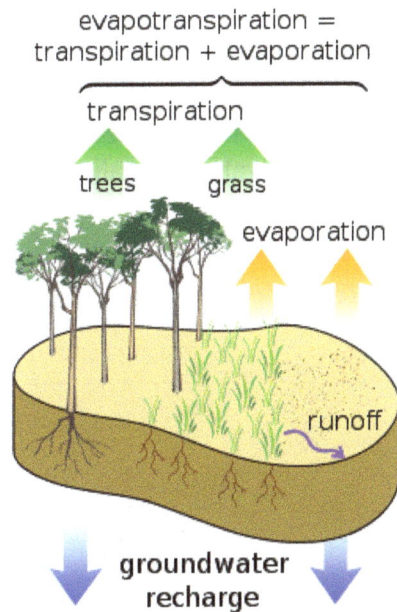

Hydrological factors at the soil surface determining the recharge

The primary coupling between groundwater and hydrological inputs is the unsaturated zone or vadose zone. The soil acts to partition hydrological inputs such as rainfall or snowmelt into surface runoff, soil moisture, evapotranspiration and groundwater recharge. Flows through the unsaturated zone that couple surface water to soil moisture and groundwater can be upward or downward, depending upon the gradient of hydraulic head in the soil, can be modeled using the numerical solution of Richards' equation partial differential equation, or the ordinary differential equation Finite Water-Content method as validated for modeling groundwater and vadose zone interactions.

Operational Inputs

The operational inputs concern human interferences with the *water management* like irrigation, drainage, pumping from wells, watertable control, and the operation of retention or infiltration basins, which are often of an hydrological nature. These inputs may also vary in time and space.

Many groundwater models are made for the purpose of assessing the effects hydraulic engineering measures.

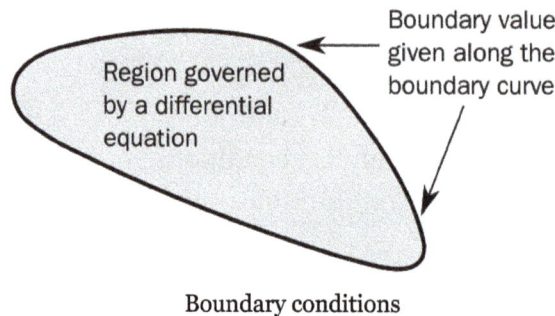

Boundary conditions

Boundary and Initial Conditions

Boundary conditions can be related to levels of the water table, artesian pressures, and hydraulic head along the boundaries of the model on the one hand (the *head conditions*), or to groundwater inflows and outflows along the boundaries of the model on the other hand (the *flow conditions*). This may also include quality aspects of the water like salinity.

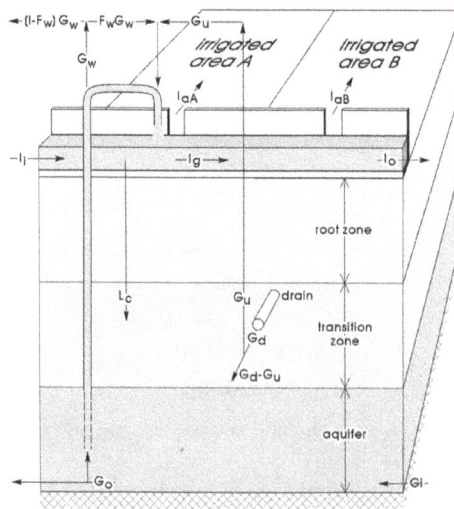

Example of parameters of an irrigation cum groundwater model

The *initial conditions* refer to initial values of elements that may increase or decrease in the course of the time *inside* the model domain and they cover largely the same phenomena as the boundary conditions do.

The initial and boundary conditions may vary from place to place. The boundary conditions may be kept either constant or be made variable in time.

Parameters

The parameters usually concern the geometry of and distances in the domain to be modelled and those physical properties of the aquifer that are more or less constant with time but that may be variable in space.

Important parameters are the topography, thicknesses of soil / rock layers and their horizontal/ vertical hydraulic conductivity (permeability for water), aquifer transmissivity and resistance, aquifer porosity and storage coefficient, as well as the capillarity of the unsaturated zone.

Some parameters may be influenced by changes in the groundwater situation, like the thickness of a soil layer that may reduce when the water table drops and/the hydraulic pressure is reduced. This phenomenon is called subsidence. The thickness, in this case, is variable in time and not a parameter proper.

Applicability

The applicability of a groundwater model to a real situation depends on the accuracy of the input data and the parameters. Determination of these requires considerable study, like collection of hydrological data (rainfall, evapotranspiration, irrigation, drainage) and determination of the parameters mentioned before including pumping tests. As many parameters are quite variable in space, expert judgment is needed to arrive at representative values.

The models can also be used for the if-then analysis: if the value of a parameter is A, then what is the result, and if the value of the parameter is B instead, what is the influence? This analysis may be sufficient to obtain a rough impression of the groundwater behavior, but it can also serve to do a *sensitivity analysis* to answer the question: which factors have a great influence and which have less influence. With such information one may direct the efforts of investigation more to the influential factors.

When sufficient data have been assembled, it is possible to determine some of missing information by **calibration**. This implies that one assumes a range of values for the unknown or doubtful value of a certain parameter and one runs the model repeatedly while comparing results with known corresponding data. For example, if salinity figures of the groundwater are available and the value of hydraulic conductivity is uncertain, one assumes a range of conductivities and the selects that value of conductivity as "true" that yields salinity results close to the observed values, meaning that the groundwater flow as governed by the hydraulic conductivity is in agreemnent with the salinity conditions. This procedure is similar to the measurement of the flow in a river or canal by letting very saline water of a known salt concentration drip into the channel and measuring the resulting salt concentration downstream.

Dimensions

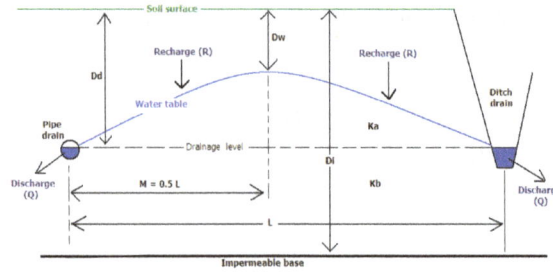

Geometry subsurface drainage system by pipes or ditches
D = depth   K = hydraulic conductivity   L = Drain spacing

Two-dimensional model of subsurface drainage in a vertical plane

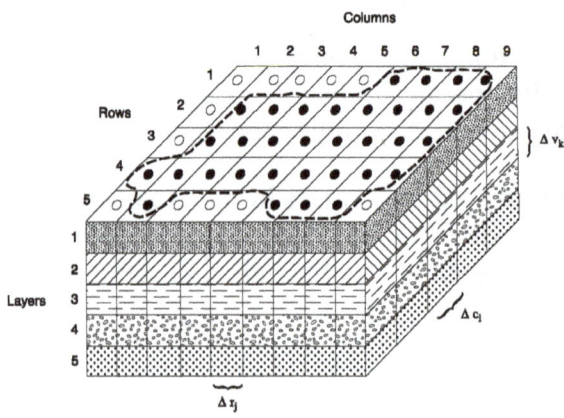

Three-dimensional grid, Modflow

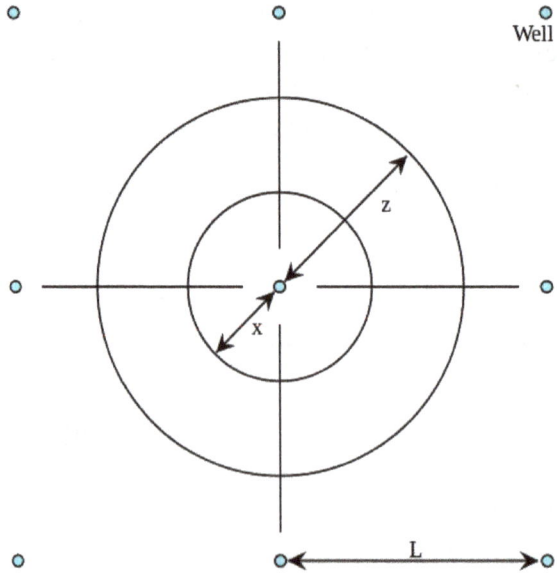

Map of a radial semi 3-dimensional model consisting of vertical concentrical cylinders through which the flow passes radially to the well

Groundwater models can be one-dimensional, two-dimensional, three-dimensional and semi-three-dimensional. Two and three-dimensional models can take into account the anisotropy of the aquifer with respect to the hydraulic conductivity, i.e. this property may vary in different directions.

One-, Two- and Three-dimensional

1. One-dimensional models can be used for the vertical flow in a system of parallel horizontal layers.

2. Two-dimensional models apply to a vertical plane while it is assumed that the groundwater conditions repeat themselves in other parallel vertical planes. Spacing equations of subsurface drains and the groundwater energy balance applied to drainage equations are examples of two-dimensional groundwater models.

3. Three-dimensional models like Modflow require discretization of the entire flow domain. To that end the flow region must be subdivided into smaller elements (or cells), in both horizontal and vertical sense. Within each cell the parameters are maintained constant, but they may vary between the cells. Using numerical solutions of groundwater flow equations, the flow of groundwater may be found as horizontal, vertical and, more often, as intermediate.

Semi three-Dimensional

In semi 3-dimensional models the horizontal flow is described by 2-dimensional flow equations (i. e. in horizontal x and y direction). Vertical flows (in z-direction) are described (a) with a 1-dimensional flow equation, or (b) derived from a water balance of horizontal flows converting the excess of horizontally incoming over the horizontally outgoing groundwater into vertical flow under the assumption that water is incompressible.

There are two classes of semi 3-dimensional models:

- *Continuous models* or *radial models* consisting of 2 dimensional submodels in vertical radial planes intersecting each other in one single axis. The flow pattern is repeated in each vertical plane fanning out from the central axis.

- *Discretized models* or *prismatic models* consisting of submodels formed by vertical blocks or prisms for the horizontal flow combined with one or more methods of superposition of the vertical flow.

Continuous Radial Model

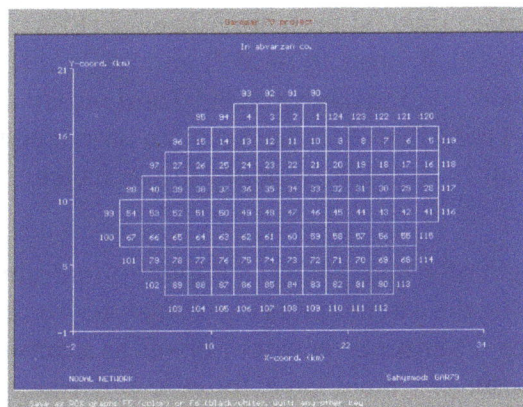

Map of a two-dimensional grid over an alluvial fan for a prismatic semi 3-dimensional model, SahysMod

Een example of a non-discretized radial model is the description of groundwater flow moving radially towards a deep well in a network of wells from which water is abstracted. The radial flow passes through a vertical, cylindrical, cross-section representing the hydraulic equipotential of which the surface diminishes in the direction of the axis of intersection of the radial planes where the well is located.

Prismatically Discretized Model

Prismatically discretized models like SahysMod have a grid over the land surface only. The 2-di-mensional grid network consists of triangles, squares, rectangles or polygons. Hence, the flow domain is subdivided into vertical blocks or prisms. The prisms can be discretized into *horizontal* layers with different characteristics that may also vary between the prisms. The ground-water flow between neighboring prisms is calculated using 2-dimensional horizontal groundwater flow equations. Vertical flows are found by applying one-dimensional flow equations in a vertical sense, or they can be derived from the water balance: excess of horizontal inflow over horizontal outflow (or vice versa) is translated into vertical flow, as demonstrated in the article Hydrology (agriculture).

In semi 3-dimensional models, intermediate flow between horizontal and vertical is not modelled like in truly 3-dimensional models. Yet, like the truly 3-dimensional models, such models do per-mit the introduction of horizontal and vertical subsurface drainage systems.

Semiconfined aquifers with a slowly permeable layer overlying the aquifer (the aquitard) can be included in the model by simulating vertical flow through it under influence of an overpressure in the aquifer proper relative to the level of the watertable inside or above the aquitard.

DSSAM Model

The DSSAM Model (Dynamic Stream Simulation and Assessment.Model) is a computer simulation developed for the Truckee River to analyze water quality impacts from land use and wastewater management decisions in the Truckee River Basin. This area includes the cities of Reno and Sparks, Nevada as well as the Lake Tahoe Basin. The model is historically and alternatively called the *Earth Metrics Truckee River Model*. Since original development in 1984-1986 under contract to the U.S. Environmental Protection Agency (EPA), the model has been refined and successive versions have been dubbed DSSAM II and DSSAM III. This hydrology transport model is based upon a pollutant loading metric called *Total maximum daily load (TMDL)*. The success of this flagship model contributed to the Agency's broadened commitment to the use of the underlying TMDL protocol in its national policy for management of most river systems in the United States.

The Truckee River has a length of over 115 miles (185 km) and drains an area of approximately 3120 square miles, not counting the extent of its Lake Tahoe sub-basin. The DSSAM model establishes numerous stations along the entire river extent as well as a considerable number of monitoring points inside the Great Basin's Pyramid Lake, the receiving waters of this closed hydrological system. Although the region is sparsely populated, it is important because Lake Tahoe is visited by 20 million persons per annum and Truckee River water quality affects at least two endangered species: the Cui-ui sucker fish and the Lahontan cutthroat trout.

Development History

Satellite photo of Pyramid Lake, September 1994

Impetus to derive a quantitative prediction model arose from a trend of historically decreasing river flow rates coupled with jurisdictional and tribal conflicts over water rights as well as concern for river biota. When expansion of the Reno-Sparks Wastewater Treatment Plant was proposed, the EPA decided to fund a large scale research effort to create simulation software and a parallel program to collect field data in the Truckee River and Pyramid Lake. For river stations water quality measurements were made in the benthic zone as well as the topic zone; in the case of Pyramid Lake boats were used to collect grab samples at varying depths and locations. Earth Metrics conducted the software development for the first generation computer model and collected field data on water quality and flow rates in the Truckee River. After model calibration, runs were made to evaluate impacts of alternative land use controls and discharge parameters for treated effluent.

The DSSAM Model is constructed to allow dynamic decay of most pollutants; for example, total nitrogen and phosphorus are allowed to be consumed by benthic algae in each time step, and the algal communities are given a separate population dynamic in each river reach (e.g.metabolic rate based upon river temperature). Sources throughout the watershed include non-point agricultural and urban stormwater as well as a multiplicity of point source discharges of treated municipal wastewater effluent.

Subsequent to the first generation of DSSAM model development, calibration and application, later refinements were made. These augmentations to model functionality focussed on increased flexibility in modeling the diel cycle and also allowed inclusion of analyzing particulate nitrogen and phosphorus. In developing DSSAM III several changes in the model operation and scope were performed.

Applications

Spawning Cui-ui sucker fish

Numerous different uses of the model have been made including (a)analysis of public policies for urban stormwater runoff, (b) researching agricultural methods for surface runoff minimization, (c) innovative solutions for non-point source control and d)engineering aspects of treated wastewater discharge. Regarding stormwater runoff in Washoe County, the specific elements within a new xeriscape ordinance were analyzed for efficacy using the model. For the varied agricultural uses in the watershed, the model was run to understand the principal sources of adverse impact, and management practices were developed to reduce in river pollution. Use of the model has specifically been conducted to analyze survival of two endangered species found in the Truckee River and Pyramid Lake: the Cui-ui sucker fish (endangered 1967) and the Lahontan cutthroat trout (threatened 1970). When the model is used for surface runoff reaching a stream, this pollutant input can be viewed as a line source (e.g., a continuous linear source of pollution entering the waterway).

Storm Water Management Model

The United States Environmental Protection Agency (EPA) Storm Water Management Model (SWMM) is a dynamic rainfall–runoff–subsurface runoff simulation model used for single-event to long-term (continuous) simulation of the surface/subsurface hydrology quantity and quality from primarily urban/suburban areas. The hydrology component of SWMM operates on a collection of subcatchment areas divided into impervious and pervious areas with and without depression storage to predict runoff and pollutant loads from precipitation, evaporation and infiltration losses from each of the subcatchment. In addition low impact development (LID) and best management practice areas on the subcatchment can be modeled to reduce the impervious and pervious runoff. The routing or hydraulics section of SWMM transports this water and possible associated water quality constituents through a system of closed pipes, open channels, storage/ treatment devices, ponds, storages, pumps, orifices, weirs, outlets, outfalls and other regulators. SWMM tracks the quantity and quality of the flow generated within each subcatchment, and the flow rate, flow depth, and quality of water in each pipe and channel during a simulation period composed of multiple fixed or variable time steps. The water quality constituents such as water quality constituents can be simulated from buildup on the subcatchments through washoff to a hydraulic network with optional first order decay and linked pollutant removal, best management practice and low-impact development (LID) removal and treatment can be simulated at selected storage nodes. SWMM is one of the hydrology transport models which the EPA and other agencies have applied widely throughout North America and through consultants and universities throughout the world. The latest update notes and new features can be found on the EPA website in the download section. Recently added in November 2015 was the EPA SWMM 5.1 Hydrology Manual

SWMM 5 Model Simulation GUI

Program Description

The EPA storm water management model (SWMM) is a dynamic rainfall-runoff-routing simulation model used for single event or long-term (continuous) simulation of runoff quantity and quality from primarily urban areas. The runoff component of SWMM operates on a collection of sub-catchment areas that receive precipitation and generate runoff and pollutant loads. The routing portion of SWMM transports this runoff through a system of pipes, channels, storage/treatment devices, pumps, and regulators. SWMM tracks the quantity and quality of runoff generated within each subcatchment, and the flow rate, flow depth, and quality of water in each pipe and channel during a simulation period divided into multiple time steps.

SWMM accounts for various hydrologic processes that produce runoff from urban areas. These include:

1. time-varying rainfall
2. evaporation of standing surface water
3. snow accumulation and melting
4. rainfall interception from depression storage
5. infiltration of rainfall into unsaturated soil layers
6. percolation of infiltrated water into groundwater layers
7. interflow between groundwater and the drainage system
8. nonlinear reservoir routing of overland flow

9. capture and retention of rainfall/runoff with various types of low impact development (LID) practices.

SWMM also contains a flexible set of hydraulic modeling capabilities used to route runoff and external inflows through the drainage system network of pipes, channels, storage/treatment units and diversion structures. These include the ability to:

1. handle networks of unlimited size·

2. use a wide variety of standard closed and open conduit shapes as well as natural channels·

3. model special elements such as storage/treatment units, flow dividers, pumps, weirs, and orifices·

4. apply external flows and water quality inputs from surface runoff, groundwater interflow, rainfall-dependent infiltration/inflow, dry weather sanitary flow, and user-defined inflows

5. utilize either kinematic wave or full dynamic wave flow routing methods·

6. model various flow regimes, such as backwater, surcharging, reverse flow, and surface ponding·

7. apply user-defined dynamic control rules to simulate the operation of pumps, orifice openings, and weir crest levels.

Spatial variability in all of these processes is achieved by dividing a study area into a collection of smaller, homogeneous subcatchment areas, each containing its own fraction of pervious and impervious sub-areas. Overland flow can be routed between sub-areas, between subcatchments, or between entry points of a drainage system.

Since its inception, SWMM has been used in thousands of sewer and stormwater studies throughout the world. Typical applications include:

1. design and sizing of drainage system components for flood control

2. sizing of detention facilities and their appurtenances for flood control and water quality protection·

3. flood plain mapping of natural channel systems, by modeling the river hydraulics and associated flooding problems using prismatic channels·

4. designing control strategies for minimizing Combined Sewer Overflow (CSO) and Sanitary Sewer Overflow (SSO)·

5. evaluating the impact of inflow and infiltration on sanitary sewer overflows·

6. generating non-point source pollutant loadings for waste load allocation studies·

7. evaluating the effectiveness of BMPs and Subcatchment LID's for reducing wet weather pollutant loadings.Rainfall-runoff modeling of urban and rural watersheds

8. hydraulic and water quality analysis of storm, sanitary, and combined sewer systems

9. master planning of sewer collection systems and urban watersheds

10. system evaluations associated with USEPA's regulations including NDPES permits, CMOM, and TMDL

11. 1D and 2D (surface ponding) predictions of flood levels and flooding volume

EPA SWMM is public domain software that may be freely copied and distributed. The SWMM 5 public domain consists of C engine code and Delphi SWMM 5 graphical user interface code. The C code and Delphi code are easily edited and can be recompiled by students and professionals for custom features or extra output features.

History

SWMM was first developed between 1969–1971 and has undergone four major upgrades since those years. The major upgrades were: (1) Version 2 in 1973-1975, (2) Version 3 in 1979-1981, (3) Version 4 in 1985-1988 and (4) Version 5 in 2001-2004. A list of the major changes and post 2004 changes are shown in Table 1. The current SWMM edition, Version 5/5.1.010, is a complete re-write of the previous Fortran releases in the programming language C, and it can be run under Windows XP, Windows Vista, Windows 7, Windows 8, Windows 10 and also with a recomplilation under Unix. The code for SWMM5 is open source and public domain code that can be downloaded from the EPA Web Site.

Table 1. SWMM History				
Release Date	**Versions**	**Developers**	**FEMA Approval**	**LID Controls**
08/20/2015	SWMM 5.1.010	EPA	Yes	Yes
04/30/2015	SWMM 5.1.009	EPA	Yes	Yes
04/17/2015	SWMM 5.1.008	EPA	Yes	Yes
10/09/2014	SWMM 5.1.007	EPA	Yes	Yes
06/02/2014	SWMM 5.1.006	EPA	Yes	Yes
03/27/2014	SWMM 5.1.001	EPA	Yes	Yes
04/21/2011	SWMM 5.0.022	EPA	Yes	Yes
08/20/2010	SWMM 5.0.019	EPA	Yes	Yes
08/17/2005	SWMM 5.0.005	EPA, CDM	Yes	No
11/30/2004	SWMM 5.0.004	EPA, CDM	No	No
11/25/2004	SWMM 5.0.003	EPA, CDM	No	No
10/26/2004	SWMM 5.0.001	EPA, CDM	No	No
2001–2004	SWMM5	EPA, CDM	No	No
1988–2004	SWMM4	UF, OSU, CDM	No	No
1981–1988	SWMM3	UF, CDM	No	No
1975–1981	SWMM2	UF	No	No
1969–1971	SWMM1	UF, CDM, M&E	No	No

EPA SWMM 5 provides an integrated graphical environment for editing watershed input data, running hydrologic, hydraulic, real time control and water quality simulations, and viewing the results in a variety of graphical formats. These include color-coded thematic drainage area maps, time series graphs and tables, profile plots, scatter plots and statistical frequency analyses.

This latest re-write of EPA SWMM was produced by the Water Supply and Water Resources Division of the U.S. Environmental Protection Agency's National Risk Management Research Laboratory with assistance from the consulting firm of CDM Inc under a Cooperative Research and Development Agreement (CRADA). SWMM 5 is used as the computational engine for many modeling packages plus components of SWMM5 are in other modeling packages. The major modeling packages that use all or some of the SWMM5 components are shown in the Vendor section. The update history of SWMM 5 from the original SWMM 5.0.001 to the current version SWMM 5.1.007 can be found at the EPA Download in the file epaswmm5_updates.txt. SWMM 5 was approved FEMA Model Approval Page in May 2005 with this note about the versions that are approved on the FEMA Approval Page SWMM 5 Version 5.0.005 (May 2005) and up for NFIP modeling. SWMM 5 is used as the computational engine for many modeling packages and some components of SWMM5 are in other modeling packages.

SWMM Conceptual Model

SWMM conceptualizes a drainage system as a series of water and material flows between several major environmental compartments. These compartments and the SWMM objects they contain include:

The Atmosphere compartment, from which precipitation falls and pollutants are deposited onto the land surface compartment. SWMM uses Rain Gage objects to represent rainfall inputs to the system. The raingage objects can use time series, external text files or NOAA rainfall data files. The Rain Gage objects can use precipitation for thousands of years. Using the SWMM-CAT Addon to SWMM5 climate change can now be simulated using modified temperature, evaporation or rainfall.

The Land Surface compartment, which is represented through one or more Subcatchment objects. It receives precipitation from the Atmospheric compartment in the form of rain or snow; it sends outflow in the form of infiltration to the Groundwater compartment and also as surface runoff and pollutant loadings to the Transport compartment. The Low Impact Development (LID) controls are part of the Subcatchments and store, infiltrate or evaporate the runoff.

The Groundwater compartment receives Infiltration (hydrology) from the Land Surface compartment and transfers a portion of this inflow to the Transport compartment. This compartment is modeled using Aquifer objects. The connection to the Transport compartment can be either a static boundary or a dynamic depth in the channels. The links in the Transport compartment now also have seepage and evaporation.

The Transport compartment contains a network of conveyance elements (channels, pipes, pumps, and regulators) and storage/treatment units that transport water to outfalls or to treatment facilities. Inflows to this compartment can come from surface runoff, groundwater interflow, sanitary dry weather flow, or from user-defined hydrographs. The components of the Transport compartment are modeled with Node and Link objects. Not all compartments need appear in a particular SWMM model. For example, one could model just the transport compartment, using pre-defined hydrographs as inputs. If you use the kinematic wave routing then the nodes do not need to contain an outfall.

Table 2. SWMM5 Compartments			
Compartment	**Compartment**	**Compartment**	**Compartment**
GAGE	SUBCATCH	NODE	LINK
TEMP	POLLUT	LANDUSE	TIMEPATTERN
CURVE	SUBCATCH	TSERIES	CONTROL
TRANSECT	AQUIFER	UNITHYD	SNOWMELT
SHAPE	LID		

Model Parameters

The simulated model parameters for subcatchments are surface roughness, depression storage, slope, flow path length; for Infiltration: Horton: max/min rates and decay constant; Green-Ampt: hydraulic conductivity, initial moisture deficit and suction head; Curve Number: NRCS (SCS) Curve number; All: time for saturated soil to fully drain; for Conduits: Manning's roughness; for Water Quality: buildup/washoff function coefficients, first order decay coefficients, removal equations. A study area can be divided into any number of individual subcatchments, each of which drains to a single point. Study areas can range in size from a small portion of a single lots up to thousands of acres. SWMM uses hourly or more frequent rainfall data as input and can be run for single events or in continuous fashion for any number of years.

Hydrology and Hydraulics Capabilities

SWMM 5 accounts for various hydrologic processes that produce surface and subsurface runoff from urban areas. These include:

1. Time-varying rainfall for an unlimited number of raingages for both design and continuous hyetographs

2. evaporation of standing surface water on watersheds and surface ponds

3. snowfall accumulation, plowing and melting

4. rainfall interception from depression storage in both impervious and pervious areas

5. infiltration of rainfall into unsaturated soil layers

6. percolation of infiltrated water into groundwater layers

7. interflow between groundwater and pipes and ditches

8. nonlinear reservoir routing of watershed overland flow.

Spatial variability in all of these processes is achieved by dividing a study area into a collection of smaller, homogeneous watershed or subcatchment areas, each containing its own fraction of pervious and impervious sub-areas. Overland flow can be routed between sub-areas, between subcatchments, or between entry points of a drainage system.

SWMM also contains a flexible set of hydraulic modeling capabilities used to route runoff and external inflows through the drainage system network of pipes, channels, storage/treatment units and diversion structures. These include the ability to:

1. Simulate drainage networks of unlimited size

2. use a wide variety of standard closed and open conduit shapes as well as natural or irregular channels

3. model special elements such as storage/treatment units, outlets, flow dividers, pumps, weirs, and orifices

4. apply external flows and water quality inputs from surface runoff, groundwater interflow, rainfall-dependent infiltration/inflow, dry weather sanitary flow, and user-defined inflows

5. utilize either steady, kinematic wave or full dynamic wave flow routing methods

6. model various flow regimes, such as backwater, surcharging, pressure, reverse flow, and surface ponding

7. apply user-defined dynamic control rules to simulate the operation of pumps, orifice openings, and weir crest levels

Infiltration is the process of rainfall penetrating the ground surface into the unsaturated soil zone of pervious subcatchments areas. SWMM5 offers four choices for modeling infiltration:

Classical Infiltration Method

This method is based on empirical observations showing that infiltration decreases exponentially from an initial maximum rate to some minimum rate over the course of a long rainfall event. Input parameters required by this method include the maximum and minimum infiltration rates, a decay coefficient that describes how fast the rate decreases over time, and the time it takes a fully saturated soil to completely dry (used to compute the recovery of infiltration rate during dry periods).

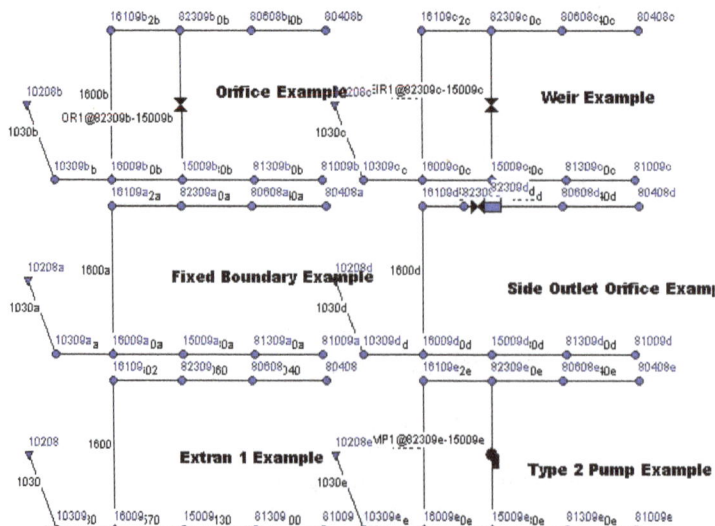

SWMM 5's QA/QC Master Example Network. This one network includes examples 1 through 7 from the SWMM 3 and SWMM 4 Manuals

Modified Horton Method

This is a modified version of the classical Horton Method that uses the cumulative infiltration in excess of the minimum rate as its state variable (instead of time along the Horton curve), providing a more accurate infiltration estimate when low rainfall intensities occur. It uses the same input parameters as does the traditional Horton Method.

Green–Ampt method

This method for modeling infiltration assumes that a sharp wetting front exists in the soil column, separating soil with some initial moisture content below from saturated soil above. The input parameters required are the initial moisture deficit of the soil, the soil's hydraulic conductivity, and the suction head at the wetting front. The recovery rate of moisture deficit during dry periods is empirically related to the hydraulic conductivity.

Curve Number Method

This approach is adopted from the NRCS (SCS) curve number method for estimating runoff. It assumes that the total infiltration capacity of a soil can be found from the soil's tabulated curve number. During a rain event this capacity is depleted as a function of cumulative rainfall and remaining capacity. The input parameters for this method are the curve number and the time it takes a fully saturated soil to completely dry (used to compute the recovery of infiltration capacity during dry periods).

SWMM also allows the infiltration recovery rate to be adjusted by a fixed amount on a monthly basis to account for seasonal variation in such factors as evaporation rates and groundwater levels. This optional monthly soil recovery pattern is specified as part of a project's evaporation data.

In addition to modeling the generation and transport of runoff flows, SWMM can also estimate the production of pollutant loads associated with this runoff. The following processes can be modeled for any number of user-defined water quality constituents:

1. Dry-weather pollutant buildup over different land uses

2. pollutant washoff from specific land uses during storm events

3. direct contribution of wet and dry rainfall deposition

4. reduction in dry-weather buildup due to street cleaning

5. reduction in washoff load due to BMPs and LIDs

6. entry of dry weather sanitary flows and user-specified external inflows at any point in the drainage system

7. routing of water quality constituents through the drainage system

8. reduction in constituent concentration through treatment in storage units or by natural processes in pipes and channels.

Rain Gages in SWMM5 supply precipitation data for one or more subcatchment areas in a study region. The rainfall data can be either a user-defined time series or come from an external file.

Several different popular rainfall file formats currently in use are supported, as well as a standard user-defined format. The principal input properties of rain gages include:

1. rainfall data type (e.g., intensity, volume, or cumulative volume)

2. recording time interval (e.g., hourly, 15-minute, etc.)

3. source of rainfall data (input time series or external file)

4. name of rainfall data source

The other principal input parameters for the subcatchments include:

1. assigned rain gage

2. outlet node or subcatchment and routing fraction

3. assigned land uses

4. tributary surface area

5. imperviousness and zero percent imperviousness

6. slope

7. characteristic width of overland flow

8. Manning's n for overland flow on both pervious and impervious areas

9. depression storage in both pervious and impervious areas

10. percent of impervious area with no depression storage.

11. infiltration parameters

12. snowpack

13. groundwater parameters

14. LID parameters for each LID Control Used

Routing Options

Steady-flow routing represents the simplest type of routing possible (actually no routing) by assuming that within each computational time step flow is uniform and steady. Thus it simply translates inflow hydrographs at the upstream end of the conduit to the downstream end, with no delay or change in shape. The normal flow equation is used to relate flow rate to flow area (or depth).

This type of routing cannot account for channel storage, backwater effects, entrance/exit losses, flow reversal or pressurized flow. It can only be used with dendritic conveyance networks, where each node has only a single outflow link (unless the node is a divider in which case two outflow links are required). This form of routing is insensitive to the time step employed and is really only appropriate for preliminary analysis using long-term continuous simulations. Kinematic wave routing solves the continuity equation along with a simplified form of the momentum equation in each conduit. The latter requires that the slope of the water surface equal the slope of the conduit.

The maximum flow that can be conveyed through a conduit is the full normal flow value. Any flow in excess of this entering the inlet node is either lost from the system or can pond atop the inlet node and be re-introduced into the conduit as capacity becomes available.

Kinematic wave routing allows flow and area to vary both spatially and temporally within a conduit. This can result in attenuated and delayed outflow hydrographs as inflow is routed through the channel. However this form of routing cannot account for backwater effects, entrance/exit losses, flow reversal, or pressurized flow, and is also restricted to dendritic network layouts. It can usually maintain numerical stability with moderately large time steps, on the order of 1 to 5 minutes. If the aforementioned effects are not expected to be significant then this alternative can be an accurate and efficient routing method, especially for long-term simulations.

Dynamic wave routing solves the complete one-dimensional Saint Venant flow equations and therefore produces the most theoretically accurate results. These equations consist of the continuity and momentum equations for conduits and a volume continuity equation at nodes.

With this form of routing it is possible to represent pressurized flow when a closed conduit becomes full, such that flows can exceed the full normal flow value. Flooding occurs when the water depth at a node exceeds the maximum available depth, and the excess flow is either lost from the system or can pond atop the node and re-enter the drainage system.

Dynamic wave routing can account for channel storage, backwater, entrance/exit losses, flow reversal, and pressurized flow. Because it couples together the solution for both water levels at nodes and flow in conduits it can be applied to any general network layout, even those containing multiple downstream diversions and loops. It is the method of choice for systems subjected to significant backwater effects due to downstream flow restrictions and with flow regulation via weirs and orifices. This generality comes at a price of having to use much smaller time steps, on the order of a minute or less (SWMM can automatically reduce the user-defined maximum time step as needed to maintain numerical stability).

Integrated Hydrology/Hydraulics

Figure 3. SWMM 5's LID processes include unlimited low-impact development or BMP objects per subcatchment and 5 types of layers.

One of the great advances in SWMM 5 was the integration of urban/suburban subsurface flow with the hydraulic computations of the drainage network. This advance is a tremendous improvement over the separate subsurface hydrologic and hydraulic computations of the previous versions of SWMM because it allows the modeler to conceptually model the same interactions that occur physically in the real open channel/shallow aquifer environment. The SWMM 5 numerical engine calculates the surface runoff, subsurface hydrology and assigns the current climate data at either the wet or dry hydrologic time step. The hydraulic calculations for the links, nodes, control rules and boundary conditions of the network are then computed at either a fixed or variable time step within the hydrologic time step by using interpolation routines and the simulated hydrologic starting and ending values. The versions of SWMM 5 greater than SWMM 5.1.007 allow the modeler to simulate climate changes by globally changing the rainfall, temperature and evaporation using monthly adjustments.

An example of this integration was the collection of the disparate SWMM 4 link types in the runoff, transport and extran blocks to one unified group of closed conduit and open channel link types in SWMM 5 and a collection of node types (Figure 2).

Low-impact Development Components

The low-impact development (LID) function was new to SWMM 5.0.019/20/21/22 and SWMM 5.1+ It is integrated within the subcatchment and allows further refinement of the overflows, infiltration flow and evaporation in rain barrel, swales, permeable paving, green roof, rain garden, bioretention and infiltration trench. The term Low-impact development (Canada/US) is used in Canada and the United States to describe a land planning and engineering design approach to managing stormwater runoff. In recent years many states in the US have adopted LID concepts and standards to enhance their approach to reducing the harmful potential for storm water pollution in new construction projects. LID takes many forms but can generally be thought of as an effort to minimize or prevent concentrated flows of storm water leaving a site. To do this the LID practice suggests that when impervious surfaces (concrete, etc.) are used, they are periodically interrupted by pervious areas which can allow the storm water to infiltrate (soak into the earth)

You can define a variety of sub processes in each LID in SWMM5 such as: surface, pavement, soil, storage, drainmat and drain.

Each type of LID has limitations on the type of sub process allowed by SWMM 5. It has a good report feature and you can have a LID summary report in the rpt file and an external report file in which you can see the surface depth, soil moisture, storage depth, surface inflow, evaporation, surface infiltration, soil percolation, storage infiltration, surface outflow and the LID continuity error. You can have multiple LID's per subcatchment and we have had no issues having many complicated LID sub networks and processes inside the Subcatchments of SWMM 5 or any continuity issues not solvable by a smaller wet hydrology time step. The types of SWMM 5 LID compartments are: storage, underdrain, surface, pavement and soil. a bio retention cell has storage, underdrain and surface compartments. an infiltration trench lid has storage, underdrain and surface compartments. A porous pavement LID has storage, underdrain and pavement compartments. A rain barrel has only storage and underdrain compartments and a vegetative swale LID has a single surface compartment. Each type of LID shares different underlying compartment objects in SWMM 5 which are called layers.

This set of equations can be solved numerically at each runoff time step to determine how an inflow hydrograph to the LID unit is converted into some combination of runoff hydrograph, sub-surface storage, sub-surface drainage, and infiltration into the surrounding native soil. In addition to Street Planters and Green Roofs, the bio-retention model just described can be used to represent Rain Gardens by eliminating the storage layer and also Porous Pavement systems by replacing the soil layer with a pavement layer.

The surface layer of the LID receives both direct rainfall and runon from other areas. It loses water through infiltration into the soil layer below it, by evapotranspiration (ET) of any water stored in depression storage and vegetative capture, and by any surface runoff that might occur. The soil layer contains an amended soil mix that can support vegetative growth. It receives infiltration from the surface layer and loses water through ET and by percolation into the storage layer below it. The storage layer consists of coarse crushed stone or gravel. It receives percolation from the soil zone above it and loses water by either infiltration into the underlying natural soil or by outflow through a perforated pipe underdrain system.

New as of July 2013, the EPA's National Stormwater Calculator is a Windows desktop application that estimates the annual amount of rainwater and frequency of runoff from a specific site anywhere in the United States. Estimates are based on local soil conditions, land cover, and historic rainfall records. The calculator accesses several national databases that provide soil, topography, rainfall, and evaporation information for the chosen site. The user supplies information about the site's land cover and selects the types of low impact development (LID) controls they would like to use on site. The LID Control features in SWMM 5.1.007 include the following among types of Green infrastructure:

StreetPlanter: Bio-retention Cells are depressions that contain vegetation grown in an engineered soil mixture placed above a gravel drainage bed. They provide storage, infiltration and evaporation of both direct rainfall and runoff captured from surrounding areas. Street planters consist of concrete boxes filled with an engineered soil that supports vegetative growth. Beneath the soil is a gravel bed that provides additional storage. The walls of a planter extend 3 to 12 inches above the soil bed to allow for ponding within the unit. The thickness of the soil growing medium ranges from 6 to 24 inches while gravel beds are 6 to 18 inches in depth. The planter's capture ratio is the ratio of its area to the impervious area whose runoff it captures.

Main Street Tree Planter, Miles City (281991376)

Raingarden:Rain Gardens are a type of bio-retention cell consisting of just the engineered soil layer with no gravel bed below it.Rain Gardens are shallow depressions filled with an engineered soil mix that supports vegetative growth. They are usually used on individual home lots to capture roof runoff. Typical soil depths range from 6 to 18 inches. The capture ratio is the ratio of the rain garden's area to the impervious area that drains onto it.

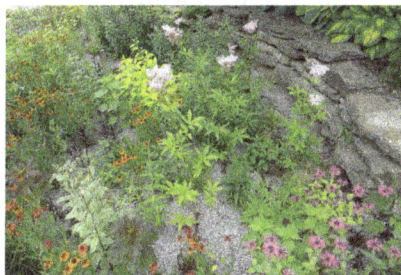

Rain garden (2014)

GreenRoof: Green Roofs are another variation of a bio-retention cell that have a soil layer laying atop a special drainage mat material that conveys excess percolated rainfall off of the roof. Green Roofs (also known as Vegetated Roofs) are bio-retention systems placed on roof surfaces that capture and temporarily store rainwater in a soil growing medium. They consist of a layered system of roofing designed to support plant growth and retain water for plant uptake while preventing ponding on the roof surface. The thickness used for the growing medium typically ranges from 3 to 6 inches.

Green roofs:

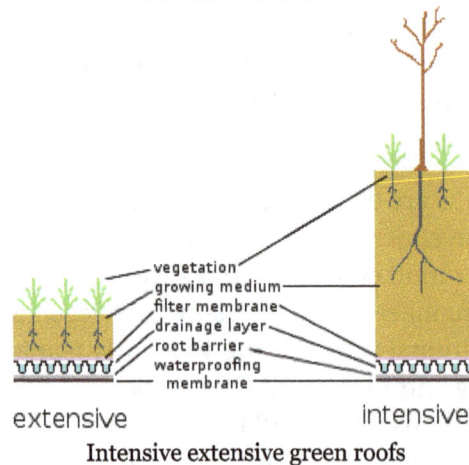

vegetation
growing medium
filter membrane
drainage layer
root barrier
waterproofing
membrane

extensive   intensive

Intensive extensive green roofs

InfilTrench: infiltration trenches are narrow ditches filled with gravel that intercept runoff from upslope impervious areas. They provide storage volume and additional time for captured runoff to infiltrate the native soil below.

Infiltration trench (6438020585)

PermPave or Permeable Pavements Continuous Permeable Pavement systems are excavated areas filled with gravel and paved over with a porous concrete or asphalt mix. Continuous Permeable Pavement systems are excavated areas filled with gravel and paved over with a porous concrete or asphalt mix. Modular Block systems are similar except that permeable block pavers are used instead. Normally all rainfall will immediately pass through the pavement into the gravel storage layer below it where it can infiltrate at natural rates into the site's native soil. Pavement layers are usually 4 to 6 inches in height while the gravel storage layer is typically 6 to 18 inches high. The Capture Ratio is the percent of the treated area (street or parking lot) that is replaced with permeable pavement.

Cistern: Rain Barrels (or Cisterns) are containers that collect roof runoff during storm events and can either release or re-use the rainwater during dry periods. Rain harvesting systems collect run-off from rooftops and convey it to a cistern tank where it can be used for non-potable water uses and on-site infiltration. The harvesting system is assumed to consist of a given number of fixed-sized cisterns per 1000 square feet of rooftop area captured. The water from each cistern is with-drawn at a constant rate and is assumed to be consumed or infiltrated entirely on-site.

VegSwale: Vegetative swales are channels or depressed areas with sloping sides covered with grass and other vegetation. They slow down the conveyance of collected runoff and allow it more time to infiltrate the native soil beneath it. Infiltration basins are shallow depressions filled with grass or other natural vegetation that capture runoff from adjoining areas and allow it to infiltrate into the soil.

Wet ponds are frequently used for water quality improvement, groundwater recharge, flood pro-tection, aesthetic improvement or any combination of these. Sometimes they act as a replace-ment for the natural absorption of a forest or other natural process that was lost when an area is developed. As such, these structures are designed to blend into neighborhoods and viewed as an amenity.

Dry ponds temporarily stores water after a storm, but eventually empties out at a controlled rate to a downstream water body.

Sand filters generally control runoff water quality, providing very limited flow rate control. A typ-ical sand filter system consists of two or three chambers or basins. The first is the sedimentation chamber, which removes floatables and heavy sediments. The second is the filtration chamber, which removes additional pollutants by filtering the runoff through a sand bed. The third is the discharge chamber. Infiltration trench, is a type of best management practice (BMP) that is used to manage stormwater runoff, prevent flooding and downstream erosion, and improve water qual-ity in an adjacent river, stream, lake or bay. It is a shallow excavated trench filled with gravel or crushed stone that is designed to infiltrate stormwater though permeable soils into the groundwa-ter aquifer.

A Vegatated filter strip is a type of buffer strip that is an area of vegetation, generally narrow and long, that slows the rate of runoff, allowing sediments, organic matter, and other pollutants that are being conveyed by the water to be removed by settling out. Filter strips reduce erosion and the accompanying stream pollution, and can be a best management practice.

Other LID like concepts around the world include sustainable drainage system (SUDS). The idea behind SUDS is to try to replicate natural systems that use cost effective solutions with low envi-ronmental impact to drain away dirty and surface water run-off through collection, storage, and cleaning before allowing it to be released slowly back into the environment, such as into water courses.

In addition the following features can also be simulated using the features of SWMM 5 (storage ponds, seepage, orifices, Weirs, seepage and evaporation from natural channels): constructed wet-lands, wet ponds, dry ponds, infiltration basin, non-surface sand filters, vegetated filterstrips, veg-etated filterstrip and infiltration basin. A WetPark would be a combination of wet and dry ponds and LID features. A WetPark is also considered a constructed wetland.

SWMM5 Components

The SWMM 5.0.001 to 5.1.010 main components are: rain gages, watersheds, LID controls or BMP features such as Wet and Dry Ponds, nodes, links, pollutants, landuses, time patterns, curves, time series, controls, transects, aquifers, unit hydrographs, snowmelt and shapes (Table 3). Other related objects are the types of Nodes and the Link Shapes. The purpose of the objects is to simulate the major components of the hydrologic cycle, the hydraulic components of the drainage, sewer or stormwater network and the buildup/washoff functions that allow the simulation of water quality constituents. A watershed simulation starts with a precipitation time history. SWMM 5 has many types of open and closed pipes and channels: dummy, circular, filled circular, rectangular closed, rectangular open, trapezoidal, triangular, parabolic, power function, rectangular triangle, rectangle round, modified baskethandle, horizontal ellipse, vertical ellipse, arch, eggshaped, horseshoe, gothic, catenary, semielliptical, baskethandle, semicircular, irregular, custom and force main.

The major objects or hydrology and hydraulic components in SWMM 5 are:

1. GAGE rain gage

2. SUBCATCH subcatchment

3. NODE conveyance system node

4. LINK conveyance system link

5. POLLUT pollutant

6. LANDUSE land use category

7. TIMEPATTERN,dry weather flow time pattern

8. CURVE generic table of values

9. TSERIES generic time series of values

10. CONTROL conveyance system control rules

11. TRANSECT irregular channel cross-section

12. AQUIFER groundwater aquifer

13. UNITHYD RDII unit hydrograph

14. SNOWMELT snowmelt parameter set

15. SHAPE custom conduit shape

16. LID LID treatment units

The major overall components are called in the SWMM 5 input file and C code of the simulation engine: gage, subcatch, node, link, pollut, landuse, timepattern, curve, tseries, control, transect, aquifer, unithyd, snowmelt, shape and lid. The subsets of possible nodes are: junction, outfall, storage and divider. Storage Nodes are either tabular with a depth/area table or a functional relationship between area and depth. Possible node inflows include: external_inflow, dry_weather_inflow, wet_weather_inflow, groundwater_inflow, rdii_inflow, flow_inflow, concen_inflow, and mass_inflow. The dry weather inflows can include the possible patterns: monthly_pattern, daily_pattern, hourly_pattern and weekend_pattern.

The SWMM 5 component structure allows the user to choose which major hydrology and hydraulic components are using during the simulation:

1. Rainfall/runoff with infiltration options: horton, modified horton, green ampt and curve number

2. RDII

3. Water Quality

4. Groundwater

5. Snowmelt

6. Flow Routing with Routing Options: Steady State, Kinematic Wave and Dynamic Wave

SWMM 3,4 to 5 Converter

The SWMM 3 and SWMM 4 converter can convert up to two files from the earlier SWMM 3 and 4 versions at one time to SWMM 5. Typically you would convert a Runoff and Transport file to SWMM 5 or a Runoff and Extran File to SWMM 5. If you have a combination of a SWMM 4 Runoff, Transport and Extran network then you will have to convert it in pieces and copy and paste the two data sets together to make one SWMM 5 data set. The x,y coordinate file is only necessary if you do not have existing x, y coordinates on the D1 line of the SWMM 4 Extran input data[set. You can use the command File=>Define Ini File to define the location of the ini file. The ini file will save your conversion project input data files and directories.

The SWMMM3 and SWMM 3.5 files are fixed format. The SWMM 4 files are free format. The converter will detect which version of SWMM is being used. The converted files can be combined using a text editor to merge the created inp files.

SWMM-CAT Climate Change AddOn

The Storm Water Management Model Climate Adjustment Tool (SWMM-CAT) is a new addition to SWMM5 (December 2014). It is a simple to use software utility that allows future climate change projections to be incorporated into the Storm Water Management Model (SWMM). SWMM was recently updated to accept a set of monthly adjustment factors for each of these time series that could represent the impact of future changes in climatic conditions. SWMM-CAT provides a set of location-specific adjustments that derived from global climate change models run as part of the World Climate Research Programme (WCRP) Coupled Model Intercomparison Project Phase 3 (CMIP3) archive (Figure 4). SWMM-CAT is a utility that adds location-specific climate change adjustments to a Storm Water Management Model (SWMM) project file. Adjustments can be applied on a monthly basis to air temperature, evaporation rates, and precipitation, as well as to the 24-hour design storm at different recurrence intervals. The source of these adjustments are global climate change models run as part of the World Climate Research Programme (WCRP) Coupled Model Intercomparison Project Phase 3 (CMIP3) archive. Downscaled results from this archive were generated and converted into changes with respect to his-torical values by USEPA's CREAT project.

Table 3. SWMM5 Input File Data Section Names				
Input Section	**Input Section**	**Input Section**	**Input Section**	**Input Section**
TITLE	OPTION	FILE	RAINGAGE	TEMP
EVAP	SUBCATCH	SUBAREA	INFIL	AQUIFER
GROUNDWATER	SNOWMELT	JUNCTION	OUTFALL	STORAGE
DIVIDER	CONDUIT	PUMP	ORIFICE	WEIR
OUTLET	XSECTION	TRANSECT	LOSSES	CONTROL
		POLLUTANT	LANDUSE	BUILDUP
WASHOFF	COVERAGE	INFLOW	DWF	PATTERN
RDII	UNITHYD	LOADING	TREATMENT	CURVE
TIMESERIES	REPORT	COORDINATE	VERTICES	POLYGON
LABEL	SYMBOL	BACKDROP	TAG	PROFILE
MAP	LID_CONTROL	LID_USAGE	GWF	ADJUST

The following steps are used to select a set of adjustments to apply to SWMM5:

1) Enter the latitude and longitude coordinates of your location if available or its 5-digit zip code. SWMM-CAT will display a range of climate change outcomes for the CMIP3 results closest to your location.

2) Select whether to use climate change projections based on either a near term or far term pro-jection period. The displayed climate change outcomes will be updated to reflect your choice.

3) Select a climate change outcome to save to SWMM. There are three choices that span the range of outcomes produced by the different global climate models used in the CMIP3 proj-ect. The Hot/Dry outcome represents a model whose average temperature change was on the high end and whose average rainfall change was on the lower end of all model projections. The Warm/Wet outcome represents a model whose average temperature change was on the lower end and whose average rainfall change was on the wetter end of the spectrum. The Median outcome is for a model whose temperature and rainfall changes were closest to the median of all models.

4) Click the Save Adjustments to SWMM link to bring up a dialog form that will allow you select an existing SWMM project file to save your adjustments to. The form will also allow you to se-lect which type of adjustments (monthly temperature, evaporation, rainfall, or 24-hour design storm) to save. Conversion of temperature and evaporation units is automatically handled depending on the unit system (US or SI) detected in the SWMM file.

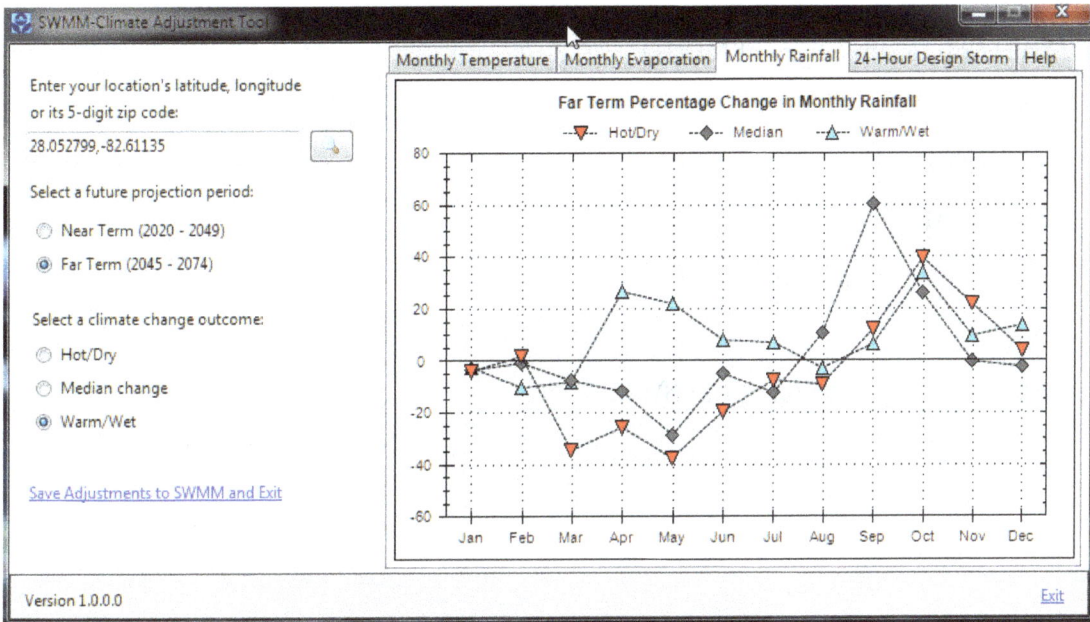

The EPA SWMM5 Climate Change Program

EPA Stormwater Calculator Based on SWMM5

Other external programs that aid in the generation of data for the EPA SWMM 5 model include: SUSTAIN, BASINS, SSOAP and the EPA's National Stormwater Calculator (SWC) which is a desktop application that estimates the annual amount of rainwater and frequency of runoff from a speci ic site anywhere in the United States (including Puerto Rico). The estimates are based on local soil conditions, land cover, and historic rainfall records.

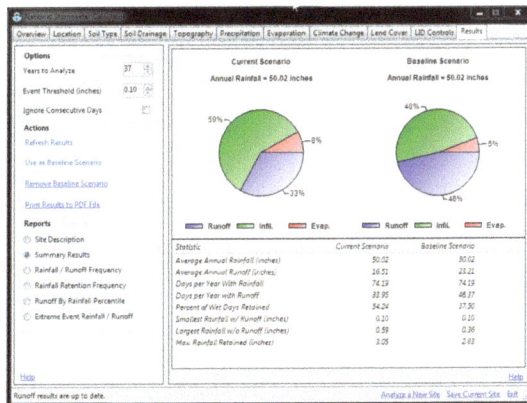

The EPA stormwater calculator for simulating long-term runoff with LID and climate change.

SWMM Platforms

A number of software packages use the SWMM5 engine, including many commercial software packages. Some of these software packages include:

- EPA-SWMM

- PCSWMM

- H2OMapSWMM, InfoSWMM, and SWMMLive, all developed by Innovyze (formerly MWH Soft, a subsidiary of MWH)

- MIKE URBAN

- XPSWMM

- GeoSWMM

- Autodesk Storm and Sanitary Analysis

- Giswater

- Free University Version of InfoSWMM

5

Various Wastewater Treatments

Wastewater treatment is the process of converting wastewater into useable water. It is water which has been harmfully affected by human influence. The various treatments discussed in the following content are industrial wastewater treatment, agricultural wastewater treatment, sewage treatment and reclaimed water. The chapter strategically encompasses and incorporates the major components and key concepts of water pollution, providing a complete understanding.

Wastewater Treatment

Wastewater treatment is a process used to convert wastewater - which is water no longer needed or suitable for its most recent use - into an effluent that can be either returned to the water cycle with minimal environmental issues or reused. The latter is called water reclamation and implies avoidance of disposal by use of treated wastewater effluent for various purposes. Treatment means removing impurities from water being treated; and some methods of treatment are applicable to both water and wastewater. The physical infrastructure used for wastewater treatment is called a "wastewater treatment plant" (WWTP).

Bundesarchiv, Bild 183-1984-1002-002
Foto: Pätzold, Ralf | 2. Oktober 1984

Clarifiers are widely used for wastewater treatment.

The treatment of wastewater belongs to the overarching field of Public Works - Environmental, with the management of human waste, solid waste, sewage treatment, stormwater (drainage) management, and water treatment. By-products from wastewater treatment plants, such as screenings, grit and sewage sludge may also be treated in a wastewater treatment plant. If the wastewater is predominantly from municipal sources (households and small industries) it is called sewage and its treatment is called sewage treatment.,

Disposal or Reuse

Although disposal or reuse occurs after treatment, it must be considered first. Since disposal or reuse are the objectives of wastewater treatment, disposal or reuse options are the basis for treatment decisions. Acceptable impurity concentrations may vary with the type of use or location of disposal. Transportation costs often make acceptable impurity concentrations dependent upon location of disposal, but expensive treatment requirements may encourage selection of a disposal location on the basis of impurity concentrations. Ocean disposal is subject to international treaty requirements. International treaties may also regulate disposal into rivers crossing international borders. Water bodies entirely within the jurisdiction of a single nation may be subject to regulations of multiple local governments. Acceptable impurity concentrations may vary widely among different jurisdictions for disposal of wastewater to evaporation ponds, infiltration basins, or injection wells.

Processes Used

Phase Separation

Phase separation transfers impurities into a non-aqueous phase. Phase separation may occur at intermediate points in a treatment sequence to remove solids generated during oxidation or polishing. Grease and oil may be recovered for fuel or saponification. Solids often require dewatering of sludge in a wastewater treatment plant. Disposal options for dried solids vary with the type and concentration of impurities removed from water.

Production of waste brine, however, may discourage wastewater treatment removing dissolved inorganic solids from water by methods like ion exchange, reverse osmosis, and distillation.

Primary settling tank of wastewater treatment plant in Dresden-Kaditz, Germany

Sedimentation

Solids and non-polar liquids may be removed from wastewater by gravity when density differences are sufficient to overcome dispersion by turbulence. Gravity separation of solids is the primary treatment of sewage, where the unit process is called "primary settling tanks" or "primary sedimentation tanks". It is also widely used for the treatment of other wastewaters. Solids that are heavier than water will accumulate at the bottom of quiescent settling basins. More complex clarifiers also have skimmers to simultaneously remove floating grease like soap scum and solids like feathers or wood chips. Containers like the API oil-water separator are specifically designed to separate non-polar liquids.

Filtration

Colloidal suspensions of fine solids may be removed by filtration through fine physical barriers distinguished from coarser screens or sieves by the ability to remove particles smaller than the openings through which the water passes. Other types of water filters remove impurities by chemical or biological processes described below.

Oxidation

Oxidation reduces the biochemical oxygen demand of wastewater, and may reduce the toxicity of some impurities. Secondary treatment converts some impurities to carbon dioxide, water, and biosolids. Chemical oxidation is widely used for disinfection.

Aeration tank of an activated sludge process at the wastewater treatment plant in Dresden-Kaditz, Germany

Biochemical Oxidation

Secondary treatment by biochemical oxidation of dissolved and colloidal organic compounds is widely used in sewage treatment and is applicable to some agricultural and industrial wastewaters. Biological oxidation will preferentially remove organic compounds useful as a food supply for the treatment ecosystem. Concentration of some less digestable compounds may be reduced by cometabolism. Removal efficiency is limited by the minimum food concentration required to sustain the treatment ecosystem. Removal efficiencies of over 96% are readily attainable through proper process controls and effective maintenance of plant equipment.

Chemical Oxidation

Chemical oxidation may remove some persistent organic pollutants and concentrations remaining after biochemical oxidation. Disinfection by chemical oxidation kills bacteria and microbial pathogens by adding ozone, chlorine or hypochlorite to wastewater.

Polishing

Polishing refers to treatments made following the above methods. These treatments may also be used independently for some industrial wastewater. Chemical reduction or pH adjustment minimizes chemical reactivity of wastewater following chemical oxidation. Carbon filtering removes remaining contaminants and impurities by chemical absorption onto activated carbon. Filtration through sand (calcium carbonate) or fabric filters is the most common method used in municipal wastewater treatment.

Wastewater Treatment Plants

Wastewater treatment plants may be distinguished by the type of wastewater to be treated, i.e. whether it is sewage, industrial wastewater, agricultural wastewater or leachate.

Overview of the wastewater treatment plant of Antwerpen-Zuid, located in the south of the agglomeration of Antwerp (Belgium)

Sewage Treatment Plants

A typical municipal sewage treatment plant in an industrialized country may include primary treatment to remove solid material, secondary treatment to digest dissolved and suspended organic material as well as the nutrients nitrogen and phosphorus, and - sometimes but not always - disinfection to kill pathogenic bacteria. The sewage sludge that is produced in sewage treatment plants undergoes sludge treatment. Larger municipalities often include factories discharging industrial wastewater into the municipal sewer system. The term "sewage treatment plant" is nowadays often replaced with the term "wastewater treatment plant".

Tertiary Treatment

Tertiary treatment is a term applied to polishing methods used following a traditional sewage treatment sequence. Tertiary treatment is being increasingly applied in industrialized countries and most common technologies are micro filtration or synthetic membranes. After membrane filtration, the treated wastewater is nearly indistinguishable from waters of natural origin of drinking quality (without its minerals). Nitrates can be removed from wastewater by natural processes in wetlands but also via microbial denitrification. Ozone wastewater treatment is also growing in popularity, and requires the use of an ozone generator, which decontaminates the water as ozone bubbles percolate through the tank, but this treatment is energy intensive. Latest, and very promising treatment technology is the use aerobic granulation.

Industrial Wastewater Treatment Plants

Disposal of wastewaters from an industrial plant is a difficult and costly problem. Most petroleum refineries, chemical and petrochemical plants have onsite facilities to treat their wastewaters so that the pollutant concentrations in the treated wastewater comply with the local and/or national regulations regarding disposal of wastewaters into community treatment plants or into rivers, lakes or oceans. Constructed wetlands are being used in an increasing number of cases as they provided high quality and productive on-site treatment. Other industrial processes that produce a lot of waste-waters such as paper and pulp production has created environmental concern, leading to development of processes to recycle water use within plants before they have to be cleaned and disposed.

Industrial wastewater treatment plants are required where municipal sewage treatment plants are unavailable or cannot adequately treat specific industrial wastewaters. Industrial wastewater plants may reduce raw water costs by converting selected wastewaters to reclaimed water used for different purposes. Industrial wastewater treatment plants may reduce wastewater treatment charges collected by municipal sewage treatment plants by pre-treating wastewaters to reduce concentrations of pollutants measured to determine user fees.

Although economies of scale may favor use of a large municipal sewage treatment plant for disposal of small volumes of industrial wastewater, industrial wastewater treatment and disposal may be less expensive than correctly apportioned costs for larger volumes of industrial wastewater not requiring the conventional sewage treatment sequence of a small municipal sewage treatment plant.

An industrial wastewater treatment plant may include one or more of the following rather than the conventional primary, secondary, and disinfection sequence of sewage treatment:

- An API oil-water separator, for removing separate phase oil from wastewater.

- A clarifier, for removing solids from wastewater.

- A roughing filter, to reduce the biochemical oxygen demand of wastewater.

- A carbon filtration plant, to remove toxic dissolved organic compounds from wastewater.

- An advanced electrodialysis reversal (EDR) system with ion exchange membranes.

Agricultural Wastewater Treatment Plants

Agricultural wastewater treatment for continuous confined animal operations like milk and egg production may be performed in plants using mechanized treatment units similar to those described under industrial wastewater; but where land is available for ponds, settling basins and facultative lagoons may have lower operational costs for seasonal use conditions from breeding or harvest cycles.

Leachate Treatment Plants

Leachate treatment plants are used to treat leachate from landfills. Treatment options include: biological treatment, mechanical treatment by ultrafiltration, treatment with active carbon filters and reverse osmosis using disc tube module technology.

Industrial Wastewater Treatment

Industrial wastewater treatment covers the mechanisms and processes used to treat wastewater that is produced as a by-product of industrial or commercial activities. After treatment, the treated industrial wastewater (or effluent) may be reused or released to a sanitary sewer or to a surface water in the environment. Most industries produce some wastewater although recent trends in the developed world have been to minimise such production or recycle such wastewater within the production process. However, many industries remain dependent on processes that produce wastewaters.

Sources of Industrial Wastewater

Complex Organic Chemicals Industry

A range of industries manufacture or use complex organic chemicals. These include pesticides, pharmaceuticals, paints and dyes, petrochemicals, detergents, plastics, paper pollution, etc. Waste waters can be contaminated by feedstock materials, by-products, product material in soluble or particulate form, washing and cleaning agents, solvents and added value products such as plasticisers. Treatment facilities that do not need control of their effluent typically opt for a type of aerobic treatment, i.e. aerated lagoons.

Electric Power Plants

Fossil-fuel power stations, particularly coal-fired plants, are a major source of industrial wastewater. Many of these plants discharge wastewater with significant levels of metals such as lead, mercury, cadmium and chromium, as well as arsenic, selenium and nitrogen compounds (nitrates and nitrites). Wastewater streams include flue-gas desulfurization, fly ash, bottom ash and flue gas mercury control. Plants with air pollution controls such as wet scrubbers typically transfer the captured pollutants to the wastewater stream.

Ash ponds, a type of surface impoundment, are a widely used treatment technology at coal-fired plants. These ponds use gravity to settle out large particulates (measured as total suspended solids) from power plant wastewater. This technology does not treat dissolved pollutants. Power stations use additional technologies to control pollutants, depending on the particular wastestream in the plant. These include dry ash handling, closed-loop ash recycling, chemical precipitation, biological treatment (such as an activated sludge process), and evaporation.

Food Industry

Wastewater generated from agricultural and food operations has distinctive characteristics that set it apart from common municipal wastewater managed by public or private sewage treatment plants throughout the world: it is biodegradable and non-toxic, but has high concentrations of biochemical oxygen demand (BOD) and suspended solids (SS). The constituents of food and agriculture wastewater are often complex to predict, due to the differences in BOD and pH in effluents from vegetable, fruit, and meat products and due to the seasonal nature of food processing and post-harvesting.

Processing of food from raw materials requires large volumes of high grade water. Vegetable washing generates waters with high loads of particulate matter and some dissolved organic matter. It may also contain surfactants.

Animal slaughter and processing produces very strong organic waste from body fluids, such as blood, and gut contents. This wastewater is frequently contaminated by significant levels of antibiotics and growth hormones from the animals and by a variety of pesticides used to control external parasites.

Processing food for sale produces wastes generated from cooking which are often rich in plant organic material and may also contain salt, flavourings, colouring material and acids or alkali. Very significant quantities of oil or fats may also be present.

Iron and Steel Industry

The production of iron from its ores involves powerful reduction reactions in blast furnaces. Cooling waters are inevitably contaminated with products especially ammonia and cyanide. Production of coke from coal in coking plants also requires water cooling and the use of water in by-products separation. Contamination of waste streams includes gasification products such as benzene, naphthalene, anthracene, cyanide, ammonia, phenols, cresols together with a range of more complex organic compounds known collectively as polycyclic aromatic hydrocarbons (PAH).

The conversion of iron or steel into sheet, wire or rods requires hot and cold mechanical transformation stages frequently employing water as a lubricant and coolant. Contaminants include hydraulic oils, tallow and particulate solids. Final treatment of iron and steel products before onward sale into manufacturing includes *pickling* in strong mineral acid to remove rust and prepare the surface for tin or chromium plating or for other surface treatments such as galvanisation or painting. The two acids commonly used are hydrochloric acid and sulfuric acid. Wastewaters include acidic rinse waters together with waste acid. Although many plants operate acid recovery plants (particularly those using hydrochloric acid), where the mineral acid is boiled away from the iron salts, there remains a large volume of highly acid ferrous sulfate or ferrous chloride to be disposed of. Many steel industry wastewaters are contaminated by hydraulic oil, also known as *soluble oil*.

Mines and Quarries

The principal waste-waters associated with mines and quarries are slurries of rock particles in water. These arise from rainfall washing exposed surfaces and haul roads and also from rock washing and grading processes. Volumes of water can be very high, especially rainfall related arisings on large sites. Some specialized separation operations, such as coal washing to separate coal from native rock using density gradients, can produce wastewater contaminated by fine particulate haematite and surfactants. Oils and hydraulic oils are also common contaminants.

Mine wastewater effluent in Peru, with neutralized pH from tailing runoff.

Wastewater from metal mines and ore recovery plants are inevitably contaminated by the minerals present in the native rock formations. Following crushing and extraction of the desirable materials, undesirable materials may enter the wastewater stream. For metal mines, this can include unwanted metals such as zinc and other materials such as arsenic. Extraction of high value metals such as gold and silver may generate slimes containing very fine particles in where physical removal of contaminants becomes particularly difficult.

Additionally, the geologic formations that harbour economically valuable metals such as copper and gold very often consist of sulphide-type ores. The processing entails grinding the rock into fine particles and then extracting the desired metal(s), with the leftover rock being known as tailings. These tailings contain a combination of not only undesirable leftover metals, but also sulphide components which eventually form sulphuric acid upon the exposure to air and water that inevitably occurs when the tailings are disposed of in large impoundments. The resulting acid mine drainage, which is often rich in heavy metals (because acids dissolve metals), is one of the many environmental impacts of mining.

Nuclear Industry

The waste production from the nuclear and radio-chemicals industry is dealt with as *Radioactive waste*.

Pulp and Paper Industry

Effluent from the pulp and paper industry is generally high in suspended solids and BOD. Plants that bleach wood pulp for paper making may generate chloroform, dioxins (including 2,3,7,8-TCDD), furans, phenols and chemical oxygen demand (COD). Stand-alone paper mills using imported pulp may only require simple primary treatment, such as sedimentation or dissolved air flotation. Increased BOD or COD loadings, as well as organic pollutants, may require biological treatment such as activated sludge or upflow anaerobic sludge blanket reactors. For mills with high inorganic loadings like salt, tertiary treatments may be required, either general membrane treatments like ultrafiltration or reverse osmosis or treatments to remove specific contaminants, such as nutrients.

Industrial Oil Contamination

Industrial applications where oil enters the wastewater stream may include vehicle wash bays, workshops, fuel storage depots, transport hubs and power generation. Often the wastewater is discharged into local sewer or trade waste systems and must meet local environmental specifications. Typical contaminants can include solvents, detergents, grit. lubricants and hydrocarbons.

Water Treatment

Many industries have a need to treat water to obtain very high quality water for demanding purposes such as environmental discharge compliance. Water treatment produces organic and mineral sludges from filtration and sedimentation. Ion exchange using natural or synthetic resins removes calcium, magnesium and carbonate ions from water, typically replacing them with sodium, chloride, hydroxyl and/or other ions. Regeneration of ion exchange columns with strong acids and alkalis produces a wastewater rich in hardness ions which are readily precipitated out, especially when in admixture with other wastewater constituents.

Wool Processing

Insecticide residues in fleeces are a particular problem in treating waters generated in wool processing. Animal fats may be present in the wastewater, which if not contaminated, can be recovered for the production of tallow or further rendering.

Treatment of Industrial Wastewater

The various types of contamination of wastewater require a variety of strategies to remove the contamination.

Brine Treatment

Brine treatment involves removing dissolved salt ions from the waste stream. Although similarities to seawater or brackish water desalination exist, industrial brine treatment may contain unique combinations of dissolved ions, such as hardness ions or other metals, necessitating specific processes and equipment.

Brine treatment systems are typically optimized to either reduce the volume of the final discharge for more economic disposal (as disposal costs are often based on volume) or maximize the recovery of fresh water or salts. Brine treatment systems may also be optimized to reduce electricity consumption, chemical usage, or physical footprint.

Brine treatment is commonly encountered when treating cooling tower blowdown, produced water from steam assisted gravity drainage (SAGD), produced water from natural gas extraction such as coal seam gas, frac flowback water, acid mine or acid rock drainage, reverse osmosis reject, chlor-alkali wastewater, pulp and paper mill effluent, and waste streams from food and beverage processing.

Brine treatment technologies may include: membrane filtration processes, such as reverse osmosis; ion exchange processes such as electrodialysis or weak acid cation exchange; or evaporation processes, such as brine concentrators and crystallizers employing mechanical vapour recompression and steam.

Reverse osmosis may not be viable for brine treatment, due to the potential for fouling caused by hardness salts or organic contaminants, or damage to the reverse osmosis membranes from hydrocarbons.

Evaporation processes are the most widespread for brine treatment as they enable the highest degree of concentration, as high as solid salt. They also produce the highest purity effluent, even distillate-quality. Evaporation processes are also more tolerant of organics, hydrocarbons, or hardness salts. However, energy consumption is high and corrosion may be an issue as the prime mover is concentrated salt water. As a result, evaporation systems typically employ titanium or duplex stainless steel materials.

Brine Management

Brine management examines the broader context of brine treatment and may include consideration of government policy and regulations, corporate sustainability, environmental impact, recycling, handling and transport, containment, centralized compared to on-site treatment, avoidance and reduction, technologies, and economics. Brine management shares some issues with leachate management and more general waste management.

Solids Removal

Most solids can be removed using simple sedimentation techniques with the solids recovered as slurry or sludge. Very fine solids and solids with densities close to the density of water pose special problems. In such case filtration or ultrafiltration may be required. Although, flocculation may be used, using alum salts or the addition of polyelectrolytes.

Oils and Grease Removal

The effective removal of oils and grease is dependent on the characteristics of the oil in terms of its suspension state and droplet size, which will in turn affect the choice of separator technology.

Oil pollution in water usually comes in four states, often in combination:

- free oil - large oil droplets sitting on the surface;
- heavy oil, which sits at the bottom, often adhering to solids like dirt;
- emulsified, where the oil droplets are heavily "chopped"; and
- dissolved oil, where the droplets are fully dispersed and not visible. Emulsified oil droplets are the most common in industrial oily wastewater and are extremely difficult to separate.

The methodology for separating the oil is dependent on the oil droplet size. Larger oil droplets such as those in free oil pollution are easily removed, but as the droplets become smaller, some separator technologies perform better than others.

Most separator technologies will have an optimum range of oil droplet sizes that can be effectively treated. This is known as the "micron rating."

Analysing the oily water to determine droplet size can be performed with a video particle analyser. Alternatively, there are commonalities in industries for oil droplet sizes. Larger droplets–greater than 60 microns–are often present in wastewater in workshops, re-fuel areas and depots. Twenty to 50 micron oil droplets often are present in vehicle wash bays, meat processing and dairy manufacturing effluent and aluminium billet cooling towers. Smaller droplets in the range of 10 to 20 microns tend to occur in workshops and condensates.

Each separator technology will have its' own performance curve outlining optimum performance based on oil droplet size. the most common separators are gravity tanks or pits, API oil-water separators or plate packs, chemical treatment via DAFs, centrifuges, media filters and hydrocyclones.

API Separators

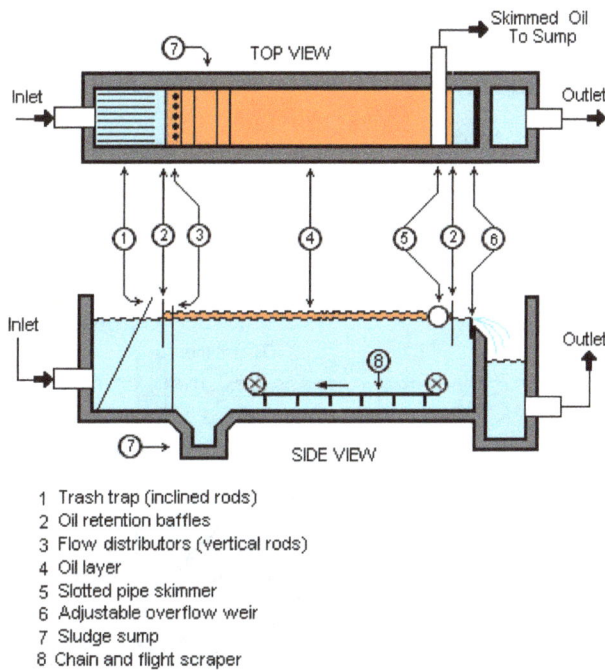

1 Trash trap (inclined rods)
2 Oil retention baffles
3 Flow distributors (vertical rods)
4 Oil layer
5 Slotted pipe skimmer
6 Adjustable overflow weir
7 Sludge sump
8 Chain and flight scraper

A typical API oil-water separator used in many industries

Many oils can be recovered from open water surfaces by skimming devices. Considered a dependable and cheap way to remove oil, grease and other hydrocarbons from water, oil skimmers can sometimes achieve the desired level of water purity. At other times, skimming is also a cost-efficient method to remove most of the oil before using membrane filters and chemical processes. Skimmers will prevent filters from blinding prematurely and keep chemical costs down because there is less oil to process.

Because grease skimming involves higher viscosity hydrocarbons, skimmers must be equipped with heaters powerful enough to keep grease fluid for discharge. If floating grease forms into solid clumps or mats, a spray bar, aerator or mechanical apparatus can be used to facilitate removal.

However, hydraulic oils and the majority of oils that have degraded to any extent will also have a soluble or emulsified component that will require further treatment to eliminate. Dissolving or emulsifying oil using surfactants or solvents usually exacerbates the problem rather than solving it, producing wastewater that is more difficult to treat.

The wastewaters from large-scale industries such as oil refineries, petrochemical plants, chemical plants, and natural gas processing plants commonly contain gross amounts of oil and suspended solids. Those industries use a device known as an API oil-water separator which is designed to separate the oil and suspended solids from their wastewater effluents. The name is derived from the fact that such separators are designed according to standards published by the American Petroleum Institute (API).

The API separator is a gravity separation device designed by using Stokes Law to define the rise velocity of oil droplets based on their density and size. The design is based on the specific gravity difference between the oil and the wastewater because that difference is much smaller than the specific gravity difference between the suspended solids and water. The suspended solids settles to the bottom of the separator as a sediment layer, the oil rises to top of the separator and the cleansed wastewater is the middle layer between the oil layer and the solids.

Typically, the oil layer is skimmed off and subsequently re-processed or disposed of, and the bottom sediment layer is removed by a chain and flight scraper (or similar device) and a sludge pump. The water layer is sent to further treatment for additional removal of any residual oil and then to some type of biological treatment unit for removal of undesirable dissolved chemical compounds.

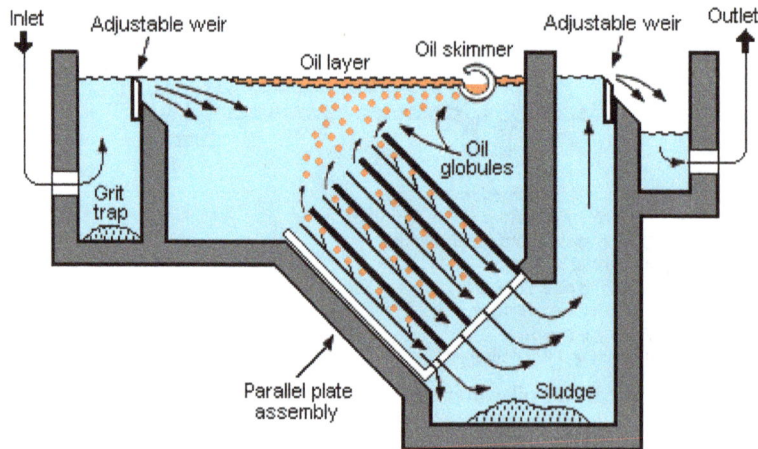

A typical parallel plate separator

Parallel plate separators are similar to API separators but they include tilted parallel plate assemblies (also known as parallel packs). The parallel plates provide more surface for suspended oil droplets to coalesce into larger globules. Such separators still depend upon the specific gravity between the suspended oil and the water. However, the parallel plates enhance the degree of oil-water separation. The result is that a parallel plate separator requires significantly less space than a conventional API separator to achieve the same degree of separation.

Hydrocyclone Oil Separators

Hydrocyclone oil separators operate on the process where wastewater enters the cyclone chamber and is spun under extreme centrifugal forces more than 1000 times the force of gravity. This force causes the water and oil droplets to separate. The separated oil is discharged from one end of the cyclone where treated water is discharged through the opposite end for further treatment, filtration or discharge.

Hydrocyclones are useful for the greatest range of oil droplet sizes operating from less than 10 microns and up and can operate continuously without water pre-treatment and at any temperature and pH. Applications where hydrocyclones are found are in industry where oily water sources arise in workshops, vehicle wash bays, transport hubs, fuel depots and aluminium billet processing. Animal fats from meat processing and dairy manufacturing can also be removed without the need of chemical treatment that often is required for dissolved air flotation (DAF) systems.

Removal of Biodegradable Organics

Biodegradable organic material of plant or animal origin is usually possible to treat using extended conventional sewage treatment processes such as activated sludge or trickling filter. Problems can arise if the wastewater is excessively diluted with washing water or is highly concentrated such as undiluted blood or milk. The presence of cleaning agents, disinfectants, pesticides, or antibiotics can have detrimental impacts on treatment processes.

Activated Sludge Process

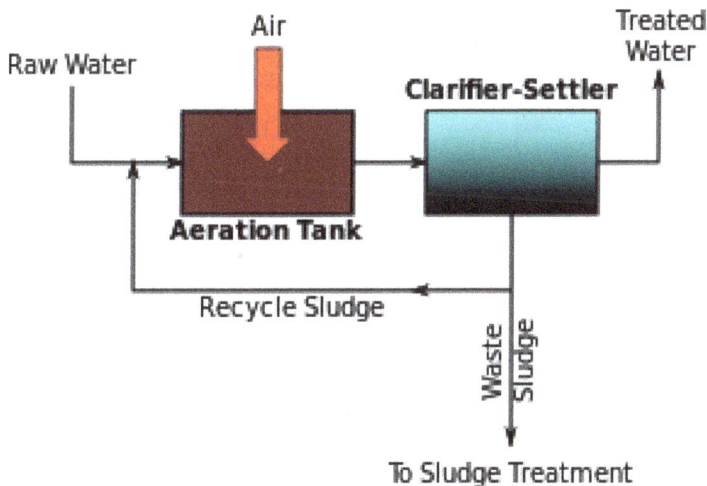

A generalized diagram of an activated sludge process.

Activated sludge is a biochemical process for treating sewage and industrial wastewater that uses air (or oxygen) and microorganisms to biologically oxidize organic pollutants, producing a waste sludge (or floc) containing the oxidized material. In general, an activated sludge process includes:

- An aeration tank where air (or oxygen) is injected and thoroughly mixed into the wastewater.

- A settling tank (usually referred to as a clarifier or "settler") to allow the waste sludge to settle. Part of the waste sludge is recycled to the aeration tank and the remaining waste sludge is removed for further treatment and ultimate disposal.

Trickling Filter Process

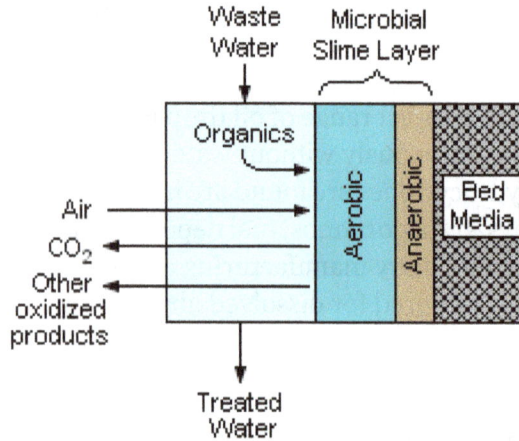

A schematic cross-section of the contact face of the bed media in a trickling filter

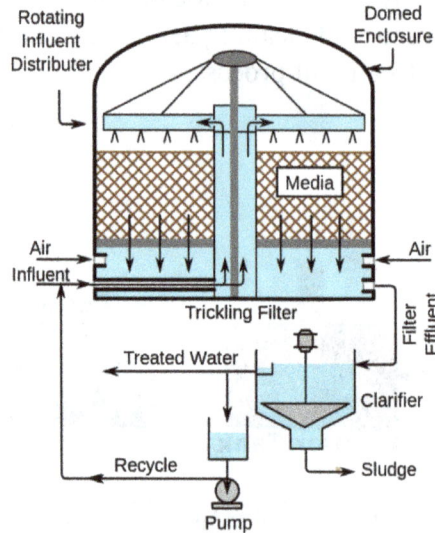

A typical complete trickling filter system

A trickling filter consists of a bed of rocks, gravel, slag, peat moss, or plastic media over which wastewater flows downward and contacts a layer (or film) of microbial slime covering the bed media. Aerobic conditions are maintained by forced air flowing through the bed or by natural convection of air. The process involves adsorption of organic compounds in the wastewater by the microbial slime layer, diffusion of air into the slime layer to provide the oxygen required for the biochemical oxidation of the organic compounds. The end products include carbon dioxide gas, water and other products of the oxidation. As the slime layer thickens, it becomes difficult for the air to penetrate the layer and an inner anaerobic layer is formed.

The fundamental components of a complete trickling filter system are:

- A bed of filter medium upon which a layer of microbial slime is promoted and developed.

- An enclosure or a container which houses the bed of filter medium.

- A system for distributing the flow of wastewater over the filter medium.

- A system for removing and disposing of any sludge from the treated effluent.

The treatment of sewage or other wastewater with trickling filters is among the oldest and most well characterized treatment technologies.

A trickling filter is also often called a *trickle filter, trickling biofilter, biofilter, biological filter* or *biological trickling filter.*

Treatment of other Organics

Synthetic organic materials including solvents, paints, pharmaceuticals, pesticides, products from coke production and so forth can be very difficult to treat. Treatment methods are often specific to the material being treated. Methods include advanced oxidation processing, distillation, adsorption, vitrification, incineration, chemical immobilisation or landfill disposal. Some materials such as some detergents may be capable of biological degradation and in such cases, a modified form of wastewater treatment can be used.

Treatment of Acids and Alkalis

Acids and alkalis can usually be neutralised under controlled conditions. Neutralisation frequently produces a precipitate that will require treatment as a solid residue that may also be toxic. In some cases, gases may be evolved requiring treatment for the gas stream. Some other forms of treatment are usually required following neutralisation.

Waste streams rich in hardness ions as from de-ionisation processes can readily lose the hardness ions in a buildup of precipitated calcium and magnesium salts. This precipitation process can cause severe *furring* of pipes and can, in extreme cases, cause the blockage of disposal pipes. A 1 metre diameter industrial marine discharge pipe serving a major chemicals complex was blocked by such salts in the 1970s. Treatment is by concentration of de-ionisation waste waters and disposal to landfill or by careful pH management of the released wastewater.

Treatment of Toxic Materials

Toxic materials including many organic materials, metals (such as zinc, silver, cadmium, thallium, etc.) acids, alkalis, non-metallic elements (such as arsenic or selenium) are generally resistant to biological processes unless very dilute. Metals can often be precipitated out by changing the pH or by treatment with other chemicals. Many, however, are resistant to treatment or mitigation and may require concentration followed by landfilling or recycling. Dissolved organics can be *incinerated* within the wastewater by the advanced oxidation process.

Agricultural Wastewater Treatment

Agricultural wastewater treatment is the treatment of wastewaters produced in the course of agricultural activities. Agriculture is a highly intensified industry in many parts of the world, producing a range of wastewaters requiring a variety of treatment technologies and management practices.

Nonpoint Source Pollution

Nonpoint source pollution from farms is caused by surface runoff from fields during rain storms. Agricultural runoff is a major source of pollution, in some cases the only source, in many watersheds.

Sediment Runoff

Highly erodible soils on a farm in Iowa

Soil washed off fields is the largest source of agricultural pollution in the United States. Excess sediment causes high levels of turbidity in water bodies, which can inhibit growth of aquatic plants, clog fish gills and smother animal larvae.

Farmers may utilize erosion controls to reduce runoff flows and retain soil on their fields. Common techniques include:

- contour ploughing
- crop mulching
- crop rotation
- planting perennial crops
- installing riparian buffers.

Nutrient Runoff

Manure spreader

Nitrogen and phosphorus are key pollutants found in runoff, and they are applied to farmland in several ways, such as in the form of commercial fertilizer, animal manure, or municipal or industrial wastewater (effluent) or sludge. These chemicals may also enter runoff from crop residues, irrigation water, wildlife, and atmospheric deposition.

Farmers can develop and implement nutrient management plans to mitigate impacts on water quality by:

- mapping and documenting fields, crop types, soil types, water bodies
- developing realistic crop yield projections
- conducting soil tests and nutrient analyses of manures and/or sludges applied
- identifying other significant nutrient sources (e.g., irrigation water)
- evaluating significant field features such as highly erodible soils, subsurface drains, and shallow aquifers
- applying fertilizers, manures, and/or sludges based on realistic yield goals and using precision agriculture techniques.

Pesticides

Aerial application (crop dusting) of pesticides over a soybean field in the U.S.

Pesticides are widely used by farmers to control plant pests and enhance production, but chemical pesticides can also cause water quality problems. Pesticides may appear in surface water due to:

- direct application (e.g. aerial spraying or broadcasting over water bodies)
- runoff during rain storms
- aerial drift (from adjacent fields).

Some pesticides have also been detected in groundwater.Farmers may use Integrated Pest Management (IPM) techniques (which can include biological pest control) to maintain control over pests, reduce reliance on chemical pesticides, and protect water quality.

There are few safe ways of disposing of pesticide surpluses other than through containment in well managed landfills or by incineration. In some parts of the world, spraying on land is a permitted method of disposal.

Point Source Pollution

Farms with large livestock and poultry operations, such as factory farms, can be a major source of point source wastewater. In the United States, these facilities are called *concentrated animal feeding operations* or *confined animal feeding operations* and are being subject to increasing government regulation.

Animal Wastes

The constituents of animal wastewater typically contain

- Strong organic content — much stronger than human sewage
- High solids concentration
- High nitrate and phosphorus content
- Antibiotics
- Synthetic hormones
- Often high concentrations of parasites and their eggs
- Spores of *Cryptosporidium* (a protozoan) resistant to drinking water treatment processes
- Spores of *Giardia*
- Human pathogenic bacteria such as *Brucella* and *Salmonella*

Animal wastes from cattle can be produced as solid or semisolid manure or as a liquid slurry. The production of slurry is especially common in housed dairy cattle.

Treatment

Whilst solid manure heaps outdoors can give rise to polluting wastewaters from runoff, this type of waste is usually relatively easy to treat by containment and/or covering of the heap.

Animal slurries require special handling and are usually treated by containment in lagoons before disposal by spray or trickle application to grassland. Constructed wetlands are sometimes used to facilitate treatment of animal wastes, as are anaerobic lagoons. Excessive application or application to sodden land or insufficient land area can result in direct runoff to watercourses, with the potential for causing severe pollution. Application of slurries to land overlying aquifers can result in direct contamination or, more commonly, elevation of nitrogen levels as nitrite or nitrate.

The disposal of any wastewater containing animal waste upstream of a drinking water intake can pose serious health problems to those drinking the water because of the highly resistant spores present in many animals that are capable of causing disabling disease in humans. This risk exists even for very low-level seepage via shallow surface drains or from rainfall run-off.

Some animal slurries are treated by mixing with straws and composted at high temperature to produce a bacteriologically sterile and friable manure for soil improvement.

Piggery Waste

Hog confinement barn or piggery

Piggery waste is comparable to other animal wastes and is processed as for general animal waste, except that many piggery wastes contain elevated levels of copper that can be toxic in the natural environment. The liquid fraction of the waste is frequently separated off and re-used in the piggery to avoid the prohibitively expensive costs of disposing of copper-rich liquid. Ascarid worms and their eggs are also common in piggery waste and can infect humans if wastewater treatment is ineffective.

Silage Liquor

Fresh or wilted grass or other green crops can be made into a semi-fermented product called silage which can be stored and used as winter forage for cattle and sheep. The production of silage often involves the use of an acid conditioner such as sulfuric acid or formic acid. The process of silage making frequently produces a yellow-brown strongly smelling liquid which is very rich in simple sugars, alcohol, short-chain organic acids and silage conditioner. This liquor is one of the most polluting organic substances known. The volume of silage liquor produced is generally in proportion to the moisture content of the ensiled material.

Treatment

Silage liquor is best treated through prevention by wilting crops well before silage making. Any silage liquor that is produced can be used as part of the food for pigs. The most effective treatment is by containment in a slurry lagoon and by subsequent spreading on land following substantial dilution with slurry. Containment of silage liquor on its own can cause structural problems in concrete pits because of the acidic nature of silage liquor.

Milking Parlour (Dairy Farming) Wastes

Although milk has a deserved reputation as an important and valuable food product, its presence in wastewaters is highly polluting because of its organic strength, which can lead to very rapid de-oxygenation of receiving waters. Milking parlour wastes also contain large volumes of washdown water, some animal waste together with cleaning and disinfection chemicals.

Treatment

Milking parlour wastes are often treated in admixture with human sewage in a local sewage treatment plant. This ensures that disinfectants and cleaning agents are sufficiently diluted and amenable to treatment. Running milking wastewaters into a farm slurry lagoon is a possible option although this tends to consume lagoon capacity very quickly. Land spreading is also a treatment option.

Slaughtering Waste

Wastewater from slaughtering activities is similar to milking parlour waste although considerably stronger in its organic composition and therefore potentially much more polluting.

Vegetable Washing Water

Washing of vegetables produces large volumes of water contaminated by soil and vegetable pieces. Low levels of pesticides used to treat the vegetables may also be present together with moderate levels of disinfectants such as chlorine.

Treatment

Most vegetable washing waters are extensively recycled with the solids removed by settlement and filtration. The recovered soil can be returned to the land.

Firewater

Although few farms plan for fires, fires are nevertheless more common on farms than on many other industrial premises. Stores of pesticides, herbicides, fuel oil for farm machinery and fertilizers can all help promote fire and can all be present in environmentally lethal quantities in firewater from fire fighting at farms.

Treatment

All farm environmental management plans should allow for containment of substantial quantities of firewater and for its subsequent recovery and disposal by specialist disposal companies. The concentration and mixture of contaminants in firewater make them unsuited to any treatment method available on the farm. Even land spreading has produced severe taste and odour problems for downstream water supply companies in the past.

Sewage Treatment

Sewage treatment is the process of removing contaminants from wastewater, primarily from household sewage. It includes physical, chemical, and biological processes to remove these contaminants and produce environmentally safe treated wastewater (or treated effluent). A by-product of sewage treatment is usually a semi-solid waste or slurry, called sewage sludge, that has to undergo further treatment before being suitable for disposal or land application.

Wastewater treatment plant in Massachusetts, United States

Sewage treatment may also be referred to as wastewater treatment, although the latter is a broader term which can also be applied to purely industrial wastewater. For most cities, the sewer system will also carry a proportion of industrial effluent to the sewage treatment plant which has usually received pretreatment at the factories themselves to reduce the pollutant load. If the sewer system is a combined sewer then it will also carry urban runoff (stormwater) to the sewage treatment plant.

Terminology

The term "sewage treatment plant" (or "sewage treatment works" in some countries) is nowadays often replaced with the term "wastewater treatment plant".

Sewage can be treated close to where the sewage is created, which may be called a "decentralized" system or even an "on-site" system (in septic tanks, biofilters or aerobic treatment systems). Alternatively, sewage can be collected and transported by a network of pipes and pump stations to a municipal treatment plant. This is called a "centralized" system, although the borders between decentralized and centralized can be variable. For this reason, the terms "semi-decentralized" and "semi-centralized" are also being used.

Origins of Sewage

Sewage is generated by residential, institutional, commercial and industrial establishments. It includes household waste liquid from toilets, baths, showers, kitchens, and sinks draining into sewers. In many areas, sewage also includes liquid waste from industry and commerce. The separation and draining of household waste into greywater and blackwater is becoming more common in the developed world, with treated greywater being permitted to be used for watering plants or recycled for flushing toilets.

Sewage Mixing with Rainwater

Sewage may include stormwater runoff or urban runoff. Sewerage systems capable of handling storm water are known as combined sewer systems. This design was common when urban sewerage systems were first developed, in the late 19th and early 20th centuries. Combined sewers require much larger and more expensive treatment facilities than sanitary sewers. Heavy volumes of storm runoff may overwhelm the sewage treatment system, causing a spill or overflow. Sanitary sewers are typically much smaller than combined sewers, and they are not designed to transport stormwater. Backups of raw sewage can occur if excessive infiltration/inflow (dilution by stormwater and/or groundwater) is allowed into a sanitary sewer system. Communities that have urbanized in the mid-20th century or later generally have built separate systems for sewage (sanitary sewers) and stormwater, because precipitation causes widely varying flows, reducing sewage treatment plant efficiency.

As rainfall travels over roofs and the ground, it may pick up various contaminants including soil particles and other sediment, heavy metals, organic compounds, animal waste, and oil and grease. Some jurisdictions require stormwater to receive some level of treatment before being discharged directly into waterways. Examples of treatment processes used for stormwater include retention basins, wetlands, buried vaults with various kinds of media filters, and vortex separators (to remove coarse solids).

Industrial Effluent

In highly regulated developed countries, industrial effluent usually receives at least pretreatment if not full treatment at the factories themselves to reduce the pollutant load, before discharge to the sewer. This process is called industrial wastewater treatment. The same does not apply to many developing countries where industrial effluent is more likely to enter the sewer if it exists, or even the receiving water body, without pretreatment.

Industrial wastewater may contain pollutants which cannot be removed by conventional sewage treatment. Also, variable flow of industrial waste associated with production cycles may upset the population dynamics of biological treatment units, such as the activated sludge process.

Process Steps

Overview

Sewage collection and treatment is typically subject to local, state and federal regulations and standards.

Treating wastewater has the aim to produce an effluent that will do as little harm as possible when discharged to the surrounding environment, thereby preventing pollution compared to releasing untreated wastewater into the environment.

Sewage treatment generally involves three stages, called primary, secondary and tertiary treatment.

- *Primary treatment* consists of temporarily holding the sewage in a quiescent basin where heavy solids can settle to the bottom while oil, grease and lighter solids float to the surface. The settled and floating materials are removed and the remaining liquid may be discharged or subjected to secondary treatment. Some sewage treatment plants that are connected to a combined sewer system have a bypass arrangement after the primary treatment unit. This means that during very heavy rainfall events, the secondary and tertiary treatment systems can be bypassed to protect them from hydraulic overloading, and the mixture of sewage and stormwater only receives primary treatment.

- *Secondary treatment* removes dissolved and suspended biological matter. Secondary treatment is typically performed by indigenous, water-borne micro-organisms in a managed habitat. Secondary treatment may require a separation process to remove the micro-organisms from the treated water prior to discharge or tertiary treatment.

- *Tertiary treatment* is sometimes defined as anything more than primary and secondary treatment in order to allow rejection into a highly sensitive or fragile ecosystem (estuaries, low-flow rivers, coral reefs,...). Treated water is sometimes disinfected chemically or physically (for example, by lagoons and microfiltration) prior to discharge into a stream, river, bay, lagoon or wetland, or it can be used for the irrigation of a golf course, green way or park. If it is sufficiently clean, it can also be used for groundwater recharge or agricultural purposes.

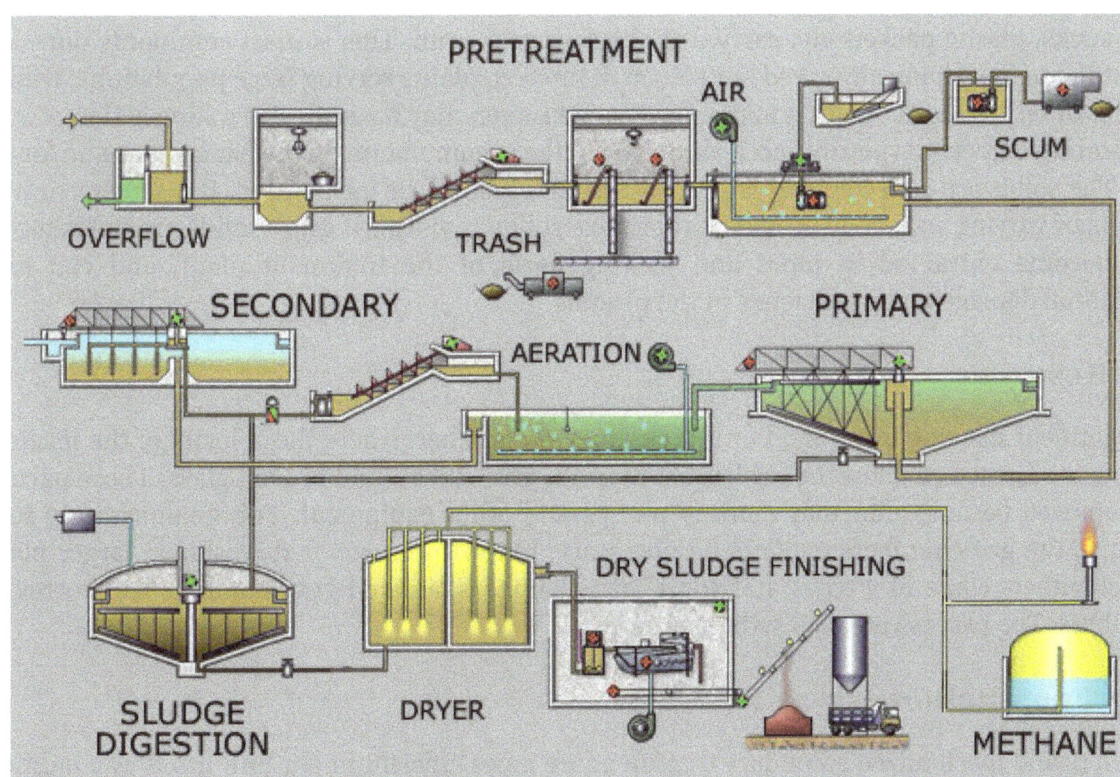

Simplified process flow diagram for a typical large-scale treatment plant

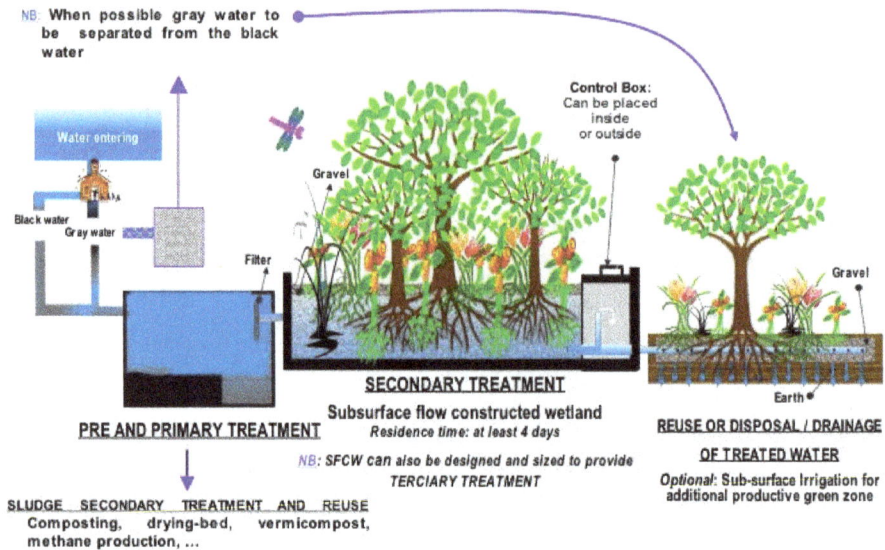

NB: When possible gray water to be separated from the black water

Water entering

Control Box: Can be placed inside or outside

Black water
Gray water
Filter

Gravel

Gravel

Earth

PRE AND PRIMARY TREATMENT

SECONDARY TREATMENT
Subsurface flow constructed wetland
Residence time: at least 4 days

NB: SFCW can also be designed and sized to provide TERCIARY TREATMENT

REUSE OR DISPOSAL / DRAINAGE OF TREATED WATER
Optional: Sub-surface Irrigation for additional productive green zone

SLUDGE SECONDARY TREATMENT AND REUSE
Composting, drying-bed, vermicompost, methane production, …

Process flow diagram for a typical treatment plant via subsurface flow constructed wetlands (SFCW)

Pretreatment

Pretreatment removes all materials that can be easily collected from the raw sewage before they damage or clog the pumps and sewage lines of primary treatment clarifiers. Objects commonly removed during pretreatment include trash, tree limbs, leaves, branches, and other large objects.

The influent in sewage water passes through a bar screen to remove all large objects like cans, rags, sticks, plastic packets etc. carried in the sewage stream. This is most commonly done with an automated mechanically raked bar screen in modern plants serving large populations, while in smaller or less modern plants, a manually cleaned screen may be used. The raking action of a mechanical bar screen is typically paced according to the accumulation on the bar screens and/or flow rate. The solids are collected and later disposed in a landfill, or incinerated. Bar screens or mesh screens of varying sizes may be used to optimize solids removal. If gross solids are not removed, they become entrained in pipes and moving parts of the treatment plant, and can cause substantial damage and inefficiency in the process.

Grit Removal

Pretreatment may include a sand or grit channel or chamber, where the velocity of the incoming sewage is adjusted to allow the settlement of sand, grit, stones, and broken glass. These particles are removed because they may damage pumps and other equipment. For small sanitary sewer systems, the grit chambers may not be necessary, but grit removal is desirable at larger plants. Grit chambers come in 3 types: horizontal grit chambers, aerated grit chambers and vortex grit chambers. The process is called sedimentation.

Flow Equalization

Clarifiers and mechanized secondary treatment are more efficient under uniform flow conditions. Equalization basins may be used for temporary storage of diurnal or wet-weather flow peaks. Basins provide a place to temporarily hold incoming sewage during plant maintenance and a means of diluting and distributing batch discharges of toxic or high-strength waste which might otherwise inhibit biological secondary treatment (including portable toilet waste, vehicle holding tanks, and septic tank pumpers). Flow equalization basins require variable discharge control, typically include provisions for bypass and cleaning, and may also include aerators. Cleaning may be easier if the basin is downstream of screening and grit removal.

Fat and Grease Removal

In some larger plants, fat and grease are removed by passing the sewage through a small tank where skimmers collect the fat floating on the surface. Air blowers in the base of the tank may also be used to help recover the fat as a froth. Many plants, however, use primary clarifiers with mechanical surface skimmers for fat and grease removal.

Primary Treatment

Primary treatment tanks in Oregon, USA.

In the primary sedimentation stage, sewage flows through large tanks, commonly called "pre-settling basins", "primary sedimentation tanks" or "primary clarifiers". The tanks are used to settle sludge while grease and oils rise to the surface and are skimmed off. Primary settling tanks are usually equipped with mechanically driven scrapers that continually drive the collected sludge towards a hopper in the base of the tank where it is pumped to sludge treatment facilities. Grease and oil from the floating material can sometimes be recovered for saponification (soap making).

Secondary Treatment

Secondary treatment is designed to substantially degrade the biological content of the sewage which are derived from human waste, food waste, soaps and detergent. The majority of municipal plants treat the settled sewage liquor using aerobic biological processes. To be effective, the biota require both oxygen and food to live. The bacteria and protozoa consume biodegradable soluble organic contaminants (e.g. sugars, fats, organic short-chain carbon molecules, etc.) and bind much of the less soluble fractions into floc. Secondary treatment systems are classified as *fixed-film* or *suspended-growth* systems.

- Fixed-film or attached growth systems include trickling filters, bio-towers, and rotating biological contactors, where the biomass grows on media and the sewage passes over its surface. The fixed-film principle has further developed into Moving Bed Biofilm Reactors (MBBR) and Integrated Fixed-Film Activated Sludge (IFAS) processes. An MBBR system typically requires a smaller footprint than suspended-growth systems.

- Suspended-growth systems include activated sludge, where the biomass is mixed with the sewage and can be operated in a smaller space than trickling filters that treat the same amount of water. However, fixed-film systems are more able to cope with drastic changes in the amount of biological material and can provide higher removal rates for organic material and suspended solids than suspended growth systems.

Secondary Sedimentation

Some secondary treatment methods include a secondary clarifier to settle out and separate biological floc or filter material grown in the secondary treatment bioreactor.

Secondary clarifier at a rural treatment plant.

List of process types

- Activated sludge
- Aerated lagoon
- Aerobic granulation
- Constructed wetland
- Membrane bioreactor
- Rotating biological contactor
- Sequencing batch reactor
- Trickling filter

To use less space, treat difficult waste, and intermittent flows, a number of designs of hybrid treatment plants have been produced. Such plants often combine at least two stages of the three main treatment stages into one combined stage. In the UK, where a large number of wastewater treatment plants serve small populations, package plants are a viable alternative to building a large structure for each process stage. In the US, package plants are typically used in rural areas, highway rest stops and trailer parks.

Tertiary Treatment

The purpose of tertiary treatment is to provide a final treatment stage to further improve the effluent quality before it is discharged to the receiving environment (sea, river, lake, wet lands, ground, etc.). More than one tertiary treatment process may be used at any treatment plant. If disinfection is practised, it is always the final process. It is also called "effluent polishing."

Filtration

Sand filtration removes much of the residual suspended matter. Filtration over activated carbon, also called *carbon adsorption,* removes residual toxins.

Lagoons or Ponds

A sewage treatment plant and lagoon in Everett, Washington, United States.

Lagoons or ponds provide settlement and further biological improvement through storage in large man-made ponds or lagoons. These lagoons are highly aerobic and colonization by native macrophytes, especially reeds, is often encouraged. Small filter feeding invertebrates such as *Daphnia* and species of *Rotifera* greatly assist in treatment by removing fine particulates.

Biological Nutrient Removal

Biological nutrient removal (BNR) is regarded by some as a type of secondary treatment process, and by others as a tertiary (or "advanced") treatment process.

Wastewater may contain high levels of the nutrients nitrogen and phosphorus. Excessive release to the environment can lead to a buildup of nutrients, called eutrophication, which can in turn encourage the overgrowth of weeds, algae, and cyanobacteria (blue-green algae). This may cause an algal bloom, a rapid growth in the population of algae. The algae numbers are unsustainable and eventually most of them die. The decomposition of the algae by bacteria uses up so much of the oxygen in the water that most or all of the animals die, which creates more organic matter for the bacteria to decompose. In addition to causing deoxygenation, some algal species produce toxins that contaminate drinking water supplies. Different treatment processes are required to remove nitrogen and phosphorus.

Nitrogen Removal

Nitrogen is removed through the biological oxidation of nitrogen from ammonia to nitrate (nitrification), followed by denitrification, the reduction of nitrate to nitrogen gas. Nitrogen gas is released to the atmosphere and thus removed from the water.

Nitrification itself is a two-step aerobic process, each step facilitated by a different type of bacteria. The oxidation of ammonia (NH_3) to nitrite (NO_2^-) is most often facilitated by *Nitrosomonas* spp. ("nitroso" referring to the formation of a nitroso functional group). Nitrite oxidation to nitrate (NO_3^-), though traditionally believed to be facilitated by *Nitrobacter* spp. (nitro referring the formation of a nitro functional group), is now known to be facilitated in the environment almost exclusively by *Nitrospira* spp.

Denitrification requires anoxic conditions to encourage the appropriate biological communities to form. It is facilitated by a wide diversity of bacteria. Sand filters, lagooning and reed beds can all be used to reduce nitrogen, but the activated sludge process (if designed well) can do the job the most easily. Since denitrification is the reduction of nitrate to dinitrogen (molecular nitrogen) gas, an electron donor is needed. This can be, depending on the waste water, organic matter (from feces), sulfide, or an added donor like methanol. The sludge in the anoxic tanks (denitrification tanks) must be mixed well (mixture of recirculated mixed liquor, return activated sludge [RAS], and raw influent) e.g. by using submersible mixers in order to achieve the desired denitrification.

Sometimes the conversion of toxic ammonia to nitrate alone is referred to as tertiary treatment.

Over time, different treatment configurations have evolved as denitrification has become more sophisticated. An initial scheme, the Ludzack-Ettinger Process, placed an anoxic treatment zone before the aeration tank and clarifier, using the return activated sludge (RAS) from the clarifier as a nitrate source. Influent wastewater (either raw or as effluent from primary clarification) serves as the electron source for the facultative bacteria to metabolize carbon, using the inorganic nitrate as a source of oxygen instead of dissolved molecular oxygen. This denitrification scheme was naturally limited to the amount of soluble nitrate present in the RAS. Nitrate reduction was limited because RAS rate is limited by the performance of the clarifier.

The "Modified Ludzak-Ettinger Process" (MLE) is an improvement on the original concept, for it recycles mixed liquor from the discharge end of the aeration tank to the head of the anoxic tank to provide a consistent source of soluble nitrate for the facultative bacteria. In this instance, raw wastewater continues to provide the electron source, and sub-surface mixing maintains the bacteria in contact with both electron source and soluble nitrate in the absence of dissolved oxygen.

Many sewage treatment plants use centrifugal pumps to transfer the nitrified mixed liquor from the aeration zone to the anoxic zone for denitrification. These pumps are often referred to as *Internal Mixed Liquor Recycle* (IMLR) pumps. IMLR may be 200% to 400% the flow rate of influent wastewater (Q.) This is in addition to Return Activated Sludge (RAS) from secondary clarifiers, which may be 100% of Q. (Therefore, the hydraulic capacity of the tanks in such a system should handle at least 400% of annual average design flow (AADF.) At times, the raw or primary effluent wastewater must be carbon-supplemented by the addition of methanol, acetate, or simple food waste (molasses, whey, plant starch) to improve the treatment efficiency. These carbon additions should be accounted for in the design of a treatment facility's organic loading.

Further modifications to the MLE were to come: Bardenpho and Biodenipho processes include additional anoxic and oxidative processes to further polish the conversion of nitrate ion to molecular nitrogen gas. Use of an anaerobic tank following the initial anoxic process allows for luxury uptake of phosphorus by bacteria, thereby biologically reducing orthophosphate ion in the treated wastewater. Even newer improvements, such as Anammox Process, interrupt the formation of nitrate at the nitrite stage of nitrification, shunting nitrite-rich mixed liquor activated sludge to treatment where nitrite is then converted to molecular nitrogen gas, saving energy, alkalinity, and secondary carbon sourcing. Anammox™ (ANaerobic AMMonia OXidation) works by artificially extending detention time and preserving denitrifiying bacteria through the use of substrate added to the mixed liquor and continuously recycled from it prior to secondary clarification. Many other proprietary schemes are being deployed, including DEMON™, Sharon-ANAMMOX™, ANITA-Mox™, and DeAmmon™. The bacteria Brocadia anammoxidans can remove ammonium from waste water through anaerobic oxidation of ammonium to hydrazine, a form of rocket fuel.

Phosphorus Removal

Every adult human excretes between 200 and 1000 grams of phosphorus annually. Studies of United States sewage in the late 1960s estimated mean per capita contributions of 500 grams in urine and feces, 1000 grams in synthetic detergents, and lesser variable amounts used as corrosion and scale control chemicals in water supplies. Source control via alternative detergent formulations has subsequently reduced the largest contribution, but the content of urine and feces will remain unchanged. Phosphorus removal is important as it is a limiting nutrient for algae growth in many fresh water systems. It is also particularly important for water reuse systems where high phosphorus concentrations may lead to fouling of downstream equipment such as reverse osmosis.

Phosphorus can be removed biologically in a process called enhanced biological phosphorus removal. In this process, specific bacteria, called polyphosphate-accumulating organisms (PAOs), are selectively enriched and accumulate large quantities of phosphorus within their cells (up to 20 percent of their mass). When the biomass enriched in these bacteria is separated from the treated water, these biosolids have a high fertilizer value.

Phosphorus removal can also be achieved by chemical precipitation, usually with salts of iron (e.g. ferric chloride), aluminum (e.g. alum), or lime. This may lead to excessive sludge production as hydroxides precipitates and the added chemicals can be expensive. Chemical phosphorus removal requires significantly smaller equipment footprint than biological removal, is easier to operate and is often more reliable than biological phosphorus removal. Another method for phosphorus removal is to use granular laterite.

Once removed, phosphorus, in the form of a phosphate-rich sewage sludge, may be dumped in a landfill or used as fertilizer. In the latter case, the treated sewage sludge is also sometimes referred to as biosolids.

Disinfection

The purpose of disinfection in the treatment of waste water is to substantially reduce the number of microorganisms in the water to be discharged back into the environment for the later use of drinking, bathing, irrigation, etc. The effectiveness of disinfection depends on the quality of the water being treated (e.g., cloudiness, pH, etc.), the type of disinfection being used, the disinfectant dosage (concentration and time), and other environmental variables. Cloudy water will be treated less successfully, since solid matter can shield organisms, especially from ultraviolet light or if contact times are low. Generally, short contact times, low doses and high flows all militate against effective disinfection. Common methods of disinfection include ozone, chlorine, ultraviolet light, or sodium hypochlorite. Chloramine, which is used for drinking water, is not used in the treatment of waste water because of its persistence. After multiple steps of disinfection, the treated water is ready to be released back into the water cycle by means of the nearest body of water or agriculture. Afterwards, the water can be transferred to reserves for everyday human uses.

Chlorination remains the most common form of waste water disinfection in North America due to its low cost and long-term history of effectiveness. One disadvantage is that chlorination of residual organic material can generate chlorinated-organic compounds that may be carcinogenic or harmful to the environment. Residual chlorine or chloramines may also be capable of chlorinating organic material in the natural aquatic environment. Further, because residual chlorine is toxic to aquatic species, the treated effluent must also be chemically dechlorinated, adding to the complexity and cost of treatment.

Ultraviolet (UV) light can be used instead of chlorine, iodine, or other chemicals. Because no chemicals are used, the treated water has no adverse effect on organisms that later consume it, as may be the case with other methods. UV radiation causes damage to the genetic structure of bacteria, viruses, and other pathogens, making them incapable of reproduction. The key disadvantages of UV disinfection are the need for frequent lamp maintenance and replacement and the need for a highly treated effluent to ensure that the target microorganisms are not shielded from the UV radiation (i.e., any solids present in the treated effluent may protect microorganisms from the UV light). In the United Kingdom, UV light is becoming the most common means of disinfection because of the concerns about the impacts of chlorine in chlorinating residual organics in the wastewater and in chlorinating organics in the receiving water. Some sewage treatment systems in Canada and the US also use UV light for their effluent water disinfection.

Ozone (O3) is generated by passing oxygen (O_2) through a high voltage potential resulting in a third oxygen atom becoming attached and forming O_3. Ozone is very unstable and reactive and oxidizes most organic material it comes in contact with, thereby destroying many pathogenic microorganisms. Ozone is considered to be safer than chlorine because, unlike chlorine which has to be stored on site (highly poisonous in the event of an accidental release), ozone is generated on-site as needed. Ozonation also produces fewer disinfection by-products than chlorination. A disadvantage of ozone disinfection is the high cost of the ozone generation equipment and the requirements for special operators.

Fourth Treatment Stage

Micropollutants such as pharmaceuticals, ingredients of household chemicals, chemicals used in small businesses or industries, environmental persistent pharmaceutical pollutant (EPPP) or pesticides may not be eliminated in the conventional treatment process (primary, secondary and tertiary treatment) and therefore lead to water pollution. Although concentrations of those substances and their decompostion products are quite low, there is still a chance to harm aquatic organisms. For pharmaceuticals, the following substances have been identified as "toxicologically relevant": substances with endocrine disrupting effects, genotoxic substances and substances that enhance the development of bacterial resistances. They mainly belong to the group of environmental persistent pharmaceutical pollutants. Techniques for elimination of micropollutants via a fourth treatment stage during sewage treatment are being tested in Germany, Switzerland and the Netherlands. However, since those techniques are still costly, they are not yet applied on a regular basis. Such process steps mainly consist of activated carbon filters that adsorb the micropollutants. Ozone can also be applied as an oxidative method. Also the use of enzymes such as the enzyme laccase is under investigation. A new concept which could provide an energy-efficient treatment of micropollutants could be the use of laccase secreting fungi cultivated at a wastewater treatment plant to degrade micropollutants and at the same time to provide enzymes at a cathode of a microbial biofuel cells. Microbial biofuel cells are investigated for their property to treat organic matter in wastewater.

To reduce pharmaceuticals in water bodies, also "source control" measures are under investigation, such as innovations in drug development or more responsible handling of drugs.

Odor Control

Odors emitted by sewage treatment are typically an indication of an anaerobic or "septic" condition. Early stages of processing will tend to produce foul-smelling gases, with hydrogen sulfide being most common in generating complaints. Large process plants in urban areas will often treat the odors with carbon reactors, a contact media with bio-slimes, small doses of chlorine, or circulating fluids to biologically capture and metabolize the noxious gases. Other methods of odor control exist, including addition of iron salts, hydrogen peroxide, calcium nitrate, etc. to manage hydrogen sulfide levels.

High-density solids pumps are suitable for reducing odors by conveying sludge through hermetic closed pipework.

Energy Requirements

For conventional sewage treatment plants, around 30 percent of the annual operating costs is usually required for energy. The energy requirements vary with type of treatment process as well as wastewater load. For example, constructed wetlands have a lower energy requirement than activated sludge plants, as less energy is required for the aeration step. Sewage treatment plants that produce biogas in their sewage sludge treatment process with anaerobic digestion can produce enough energy to meet most of the energy needs of the sewage treatment plant itself.

In conventional secondary treatment processes, most of the electricity is used for aeration, pumping systems and equipment for the dewatering and drying of sewage sludge. Advanced wastewater treatment plants, e.g. for nutrient removal, require more energy than plants that only achieve primary or secondary treatment.

Sludge Treatment and Disposal

The sludges accumulated in a wastewater treatment process must be treated and disposed of in a safe and effective manner. The purpose of digestion is to reduce the amount of organic matter and the number of disease-causing microorganisms present in the solids. The most common treatment options include anaerobic digestion, aerobic digestion, and composting. Incineration is also used, albeit to a much lesser degree.

Sludge treatment depends on the amount of solids generated and other site-specific conditions. Composting is most often applied to small-scale plants with aerobic digestion for mid-sized operations, and anaerobic digestion for the larger-scale operations.

The sludge is sometimes passed through a so-called pre-thickener which de-waters the sludge. Types of pre-thickeners include centrifugal sludge thickeners rotary drum sludge thickeners and belt filter presses. Dewatered sludge may be incinerated or transported offsite for disposal in a landfill or use as an agricultural soil amendment.

Environment Aspects

The outlet of the Karlsruhe sewage treatment plant flows into the Alb.

Many processes in a wastewater treatment plant are designed to mimic the natural treatment processes that occur in the environment, whether that environment is a natural water body or the ground. If not overloaded, bacteria in the environment will consume organic contaminants, although this will reduce the levels of oxygen in the water and may significantly change the overall ecology of the receiving water. Native bacterial populations feed on the organic contaminants, and the numbers of disease-causing microorganisms are reduced by natural environmental conditions such as predation or exposure to ultraviolet radiation. Consequently, in cases where the receiving environment provides a high level of dilution, a high degree of wastewater treatment may not be required. However, recent evidence has demonstrated that very low levels of specific contaminants in wastewater, including hormones (from animal husbandry and residue from human hormonal contraception methods) and synthetic materials such as phthalates that mimic hormones in their action, can have an unpredictable adverse impact on the natural biota and potentially on humans if the water is re-used for drinking water. In the US and EU, uncontrolled discharges of wastewater to the environment are not permitted under law, and strict water quality requirements are to be met, as clean drinking water is essential. A significant threat in the coming decades will be the increasing uncontrolled discharges of wastewater within rapidly developing countries.

Effects on Biology

Sewage treatment plants can have multiple effects on nutrient levels in the water that the treated sewage flows into. These nutrients can have large effects on the biological life in the water in contact with the effluent. Stabilization ponds (or sewage treatment ponds) can include any of the following:

- Oxidation ponds, which are aerobic bodies of water usually 1–2 meters in depth that receive effluent from sedimentation tanks or other forms of primary treatment.

  - Dominated by algae

- Polishing ponds are similar to oxidation ponds but receive effluent from an oxidation pond or from a plant with an extended mechanical treatment.

  - Dominated by zooplankton

- Facultative lagoons, raw sewage lagoons, or sewage lagoons are ponds where sewage is added with no primary treatment other than coarse screening. These ponds provide effective treatment when the surface remains aerobic; although anaerobic conditions may develop near the layer of settled sludge on the bottom of the pond.

- Anaerobic lagoons are heavily loaded ponds.

  - Dominated by bacteria

- Sludge lagoons are aerobic ponds, usually 2 to 5 meters in depth, that receive anaerobically digested primary sludge, or activated secondary sludge under water.

  - Upper layers are dominated by algae

Phosphorus limitation is a possible result from sewage treatment and results in flagellate-dominated plankton, particularly in summer and fall.

A phytoplankton study found high nutrient concentrations linked to sewage effluents. High nutrient concentration leads to high chlorophyll a concentrations, which is a proxy for primary production in marine environments. High primary production means high phytoplankton populations and most likely high zooplankton populations, because zooplankton feed on phytoplankton. However, effluent released into marine systems also leads to greater population instability.

The planktonic trends of high populations close to input of treated sewage is contrasted by the bacterial trend. In a study of *Aeromonas* spp. in increasing distance from a wastewater source, greater change in seasonal cycles was found the furthest from the effluent. This trend is so strong that the furthest location studied actually had an inversion of the *Aeromonas* spp. cycle in comparison to that of fecal coliforms. Since there is a main pattern in the cycles that occurred simultaneously at all stations it indicates seasonal factors (temperature, solar radiation, phytoplankton) control of the bacterial population. The effluent dominant species changes from *Aeromonas caviae* in winter to *Aeromonas sobria* in the spring and fall while the inflow dominant species is *Aeromonas caviae*, which is constant throughout the seasons.

Treated Sewage Reuse

With suitable technology, it is possible to reuse sewage effluent for drinking water, although this is usually only done in places with limited water supplies, such as Windhoek and Singapore.

In Israel, about 50 percent of agricultural water use (total use was 1 billion cubic metres in 2008) is provided through reclaimed sewer water. Future plans call for increased use of treated sewer water as well as more desalination plants.

Sewage Treatment in Developing Countries

Few reliable figures exist on the share of the wastewater collected in sewers that is being treated in the world. A global estimate by UNDP and UN-Habitat is that 90% of all wastewater generated is released into the environment untreated. In many developing countries the bulk of domestic and industrial wastewater is discharged without any treatment or after primary treatment only.

In Latin America about 15 percent of collected wastewater passes through treatment plants (with varying levels of actual treatment). In Venezuela, a below average country in South America with respect to wastewater treatment, 97 percent of the country's sewage is discharged raw into the environment. In Iran, a relatively developed Middle Eastern country, the majority of Tehran's population has totally untreated sewage injected to the city's groundwater. However, the construction of major parts of the sewage system, collection and treatment, in Tehran is almost complete, and under development, due to be fully completed by the end of 2012. In Isfahan, Iran's third largest city, sewage treatment was started more than 100 years ago.

Only few cities in sub-Saharan Africa have sewer-based sanitation systems, let alone wastewater treatment plants, an exception being South Africa and – until the late 1990s- Zimbabwe. Instead, most urban residents in sub-Saharan Africa rely on on-site sanitation systems without sewers, such as septic tanks and pit latrines, and faecal sludge management in these cities is an enormous challenge.

History

Basic sewer systems were used for waste removal in ancient Mesopotamia, where vertical shafts carried the waste away into cesspools. Similar systems existed in the Indus Valley civilization in modern-day India and in Ancient Crete and Greece. In the Middle Ages the sewer systems built by the Romans fell into disuse and waste was collected into cesspools that were periodically emptied by workers known as 'rakers' who would often sell it as fertilizer to farmers outside the city.

FARADAY GIVING HIS CARD TO FATHER THAMES;
And we hope the Dirty Fellow will consult the learned Professor.

The Great Stink of 1858 stimulated research into the problem of sewage treatment. In this caricature in *The Times*, Michael Faraday reports to *Father Thames* on the state of the river.

Modern sewage systems were first built in the mid-nineteenth century as a reaction to the exacerbation of sanitary conditions brought on by heavy industrialization and urbanization. Due to the contaminated water supply, cholera outbreaks occurred in 1832, 1849 and 1855 in London, killing tens of thousands of people. This, combined with the Great Stink of 1858, when the smell of untreated human waste in the River Thames became overpowering, and the report into sanitation reform of the Royal Commissioner Edwin Chadwick, led to the Metropolitan Commission of Sewers appointing Sir Joseph Bazalgette to construct a vast underground sewage system for the safe removal of waste. Contrary to Chadwick's recommendations, Bazalgette's system, and others later built in Continental Europe, did not pump the sewage onto farm land for use as fertilizer; it was simply piped to a natural waterway away from population centres, and pumped back into the environment.

Early Attempts

One of the first attempts at diverting sewage for use as a fertilizer in the farm was made by the cotton mill owner James Smith in the 1840s. He experimented with a piped distribution system initially proposed by James Vetch that collected sewage from his factory and pumped it into the outlying farms, and his success was enthusiastically followed by Edwin Chadwick and supported by organic chemist Justus von Liebig.

The idea was officially adopted by the Health of Towns Commission, and various schemes (known as sewage farms) were trialled by different municipalities over the next 50 years. At first, the heavier solids were channeled into ditches on the side of the farm and were covered over when full, but soon flat-bottomed tanks were employed as reservoirs for the sewage; the earliest patent was taken out by William Higgs in 1846 for "tanks or reservoirs in which the contents of sewers and drains from cities, towns and villages are to be collected and the solid animal or vegetable matters therein contained, solidified and dried…" Improvements to the design of the tanks included the introduction of the horizontal-flow tank in the 1850s and the radial-flow tank in 1905. These tanks had to be manually de-sludged periodically, until the introduction of automatic mechanical de-sludgers in the early 1900s.

The precursor to the modern septic tank was the cesspool in which the water was sealed off to prevent contamination and the solid waste was slowly liquified due to anaerobic action; it was invented by L.H Mouras in France in the 1860s. Donald Cameron, as City Surveyor for Exeter patented an improved version in 1895, which he called a 'septic tank'; septic having the meaning of 'bacterial'. These are still in worldwide use, especially in rural areas unconnected to large-scale sewage systems.

Chemical Treatment

It was not until the late 19th century that it became possible to treat the sewage by chemically breaking it down through the use of microorganisms and removing the pollutants. Land treatment was also steadily becoming less feasible, as cities grew and the volume of sewage produced could no longer be absorbed by the farmland on the outskirts.

Sir Edward Frankland, a distinguished chemist, who demonstrated the possibility of chemically treating sewage in the 1870s.

Sir Edward Frankland conducted experiments at the Sewage Farm in Croydon, England, during the 1870s and was able to demonstrate that filtration of sewage through porous gravel produced a nitrified effluent (the ammonia was converted into nitrate) and that the filter remained unclogged over long periods of time. This established the then revolutionary possibility of biological treatment of sewage using a contact bed to oxidize the waste. This concept was taken up by the chief chemist for the London Metropolitan Board of Works, William Libdin, in 1887:

> …in all probability the true way of purifying sewage…will be first to separate the sludge, and then turn into neutral effluent… retain it for a sufficient period, during which time it should be fully aerated, and finally discharge it into the stream in a purified condition. This is indeed what is aimed at and imperfectly accomplished on a sewage farm.

From 1885 to 1891 filters working on this principle were constructed throughout the UK and the idea was also taken up in the US at the Lawrence Experiment Station in Massachusetts, where Frankland's work was confirmed. In 1890 the LES developed a 'trickling filter' that gave a much more reliable performance.

Contact beds were developed in Salford, Lancashire and by scientists working for the London City Council in the early 1890s. According to Christopher Hamlin, this was part of a conceptual revolution that replaced the philosophy that saw "sewage purification as the prevention of decomposition with one that tried to facilitate the biological process that destroy sewage naturally."

Contact beds were tanks containing the inert substance, such as stones or slate, that maximized the surface area available for the microbial growth to break down the sewage. The sewage was held in the tank until it was fully decomposed and it was then filtered out into the ground. This method quickly became widespread, especially in the UK, where it was used in Leicester, Sheffield, Manchester and Leeds. The bacterial bed was simultaneously developed by Joseph Corbett as Borough Engineer in Salford and experiments in 1905 showed that his method was superior in that greater volumes of sewage could be purified better for longer periods of time than could be achieved by the contact bed.

The Royal Commission on Sewage Disposal published its eighth report in 1912 that set what became the international standard for sewage discharge into rivers; the '20:30 standard', which allowed 20 mg Biochemical oxygen demand and 30 mg suspended solid per litre.

Reclaimed Water

Reclaimed water or recycled water, is former wastewater (sewage) that is treated to remove solids and impurities, and used in sustainable landscaping irrigation, to recharge groundwater aquifers, to meet commercial and industrial water needs, and for drinking. The purpose of these processes is water conservation and sustainability, rather than discharging the treated water to surface waters such as rivers and oceans. In some cases, recycled water can be used for streamflow augmentation to benefit ecosystems and improve aesthetics. One example of this is along Calera Creek in the City of Pacifica, CA.

Samples of different types of (waste)water, starting with raw sewage then plant effluent and finally reclaimed water (after several treatment steps)

The definition of reclaimed water, as defined by Levine and Asano, is "The end product of waste-water reclamation that meets water quality requirements for biodegradable materials, suspended matter and pathogens." Simply stated, reclaimed water is water that is used more than one time before it passes back into the natural water cycle. Scientifically-proven advances in water technology allow communities to reuse water for many different purposes, including industrial, irrigation, and drinking. The water is treated differently depending upon the source and use of the water and how it gets delivered.

Cycled repeatedly through the planetary hydrosphere, all water on Earth is recycled water, but the terms "recycled water" or "reclaimed water" typically mean wastewater sent from a home or business through a pipeline system to a treatment facility, where it is treated to a level consistent with its intended use. The water is then routed directly to a recycled water system for uses such as irrigation or industrial cooling.

There are examples of communities that have safely used recycled water for many years. Los Angeles County's sanitation districts have provided treated wastewater for landscape irrigation in parks and golf courses since 1929. The first reclaimed water facility in California was built at San Francisco's Golden Gate Park in 1932. The Irvine Ranch Water District (IRWD) was the first water district in California to receive an unrestricted use permit from the state for its recycled water; such a permit means that water can be used for any purpose except drinking. IRWD maintains one of the largest recycled water systems in the nation with more than 400 miles serving more than 4,500 metered connections. The Irvine Ranch Water District and Orange County Water District in Southern California are established leaders in recycled water. Further, the Orange County Water District, located in Orange County, and in other locations throughout the world such as Singapore, water is given more advanced treatments and is used indirectly for drinking.

In spite of quite simple methods that incorporate the principles of water-sensitive urban design (WSUD) for easy recovery of stormwater runoff, there remains a common perception that reclaimed water must involve sophisticated and technically complex treatment systems, attempting to recover the most complex and degraded types of sewage. As this effort is driven by sustainability factors, this type of implementation should inherently be associated with point source solutions, where it is most economical to achieve the expected outcomes. Harvesting of stormwater or rainwater can be an extremely simple to comparatively complex, as well as energy and chemical intensive, recovery of more contaminated sewage.

Terminology

Effluent storage tank from where treated effluent (after constructed wetland) is pumped away for irrigation, Haran-Al-Awamied, Syria

There is no one-size-fits-all solution to water reuse, but there are many safe and scientifically-proven options that allow communities to sustain their local water supplies. Below are terms scientists and water experts use to describe some of these reclaimed water options:

Reused water is water used more than once or recycled.

Potable water is drinking water.

Potable reuse refers to reused water you can drink.

Nonpotable reuse refers to reused water that is not used for drinking, but is safe to use for irrigation or industrial purposes.

De facto, unacknowledged or unplanned potable reuse occurs when water intakes draw raw water supplies downstream from discharges of treated effluent from wastewater treatment plants/water reclamation facilities or resource recovery facilities. For example, if you are downstream of a community, that community's used water (run-off and treated wastewater) gets put back into river or stream and is delivered downstream to your community and becomes part of your drinking water supply.

Planned potable reuse is publicly acknowledged as an intentional project to recycle water for drinking water. It can be either direct or indirect. It commonly involves a more formal public process and public consultation program than is observed with de facto or unacknowledged reuse.

How potable reused water is delivered determines if it is called Indirect Potable Reuse or Direct Potable Reuse.

- Indirect potable reuse means the water is delivered to you indirectly. After it is purified, the reused water blends with other supplies and/or sits a while in some sort of storage, man-made or natural, before it gets delivered to a pipeline that leads to a water treatment plant or distribution system. That storage could be a groundwater basin or a surface water reservoir.

- Direct potable reuse means the reused water is put directly into pipelines that go to a water treatment plant or distribution system. Direct potable reuse may occur with or without "engineered storage" such as underground or above ground tanks.

Greywater uses the same waste as water reclamation with the exception of toilet water. Greywater can be used for irrigation, toilet flushing or other domestic uses.

Desalination is an energy-intensive process where salt and other minerals are removed from sea water to produce potable water for drinking and irrigation, typically through membrane filtration (reverse-osmosis), and steam-distillation.

Maximum water recovery - To determine maximum water recovery there are various techniques that have been developed by researchers; for maximum water reuse/reclamation/recovery strategies such as water pinch analysis. The techniques help a user to target the minimum freshwater consumption and wastewater target. It also helps in designing the network that achieves the target. This provides a benchmark to be used by users in improving their water systems.

Applications

Most of the uses of water reclamation are non potable uses such as: Washing Cars, flushing toilets, cooling water for power plants, concrete mixing, artificial lakes, irrigation for golf courses and public parks, and for hydraulic fracturing. Where applicable, systems run a dual piping system to keep the recycled water separate from the potable water.

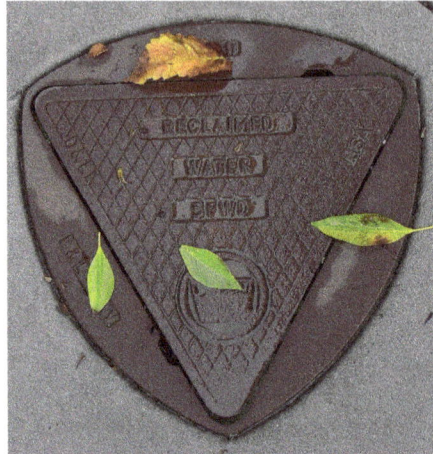

Cover for reclaimed water valve, San Francisco Water District

New technologies for recycling allow the water to be used for fracking purposes and can save an estimated 4 - 7 million or more gallons per well.

Potable uses

Some water agencies reuse highly treated effluent from municipal wastewater or resource recovery plants as a reliable, drought proof source of drinking water. By using advanced purification processes, they produce water that meets all applicable drinking water standards. System reliability and frequent monitoring and testing are imperative to them meeting stringent controls.

The water needs of a community, water sources, public health regulations, costs, and the types of water infrastructure in place, such as distribution systems, man-made reservoirs, or natural groundwater basins, determine if and how reclaimed water can be part of the drinking water supply. Communities in El Paso, Texas and Orange County, California, for example, reuse water to replenish groundwater basins. Others, such as the Upper Occoquan Service Authority in Virginia, put it into surface water reservoirs. In these instances the reclaimed water is blended with other water supplies and/or sits in storage for a certain amount of time before it is drawn out and gets treated again at a water treatment or distribution system. In some Texas communities, the reused water is put directly into pipelines that go to a water treatment plant or distribution system. In Singapore reclaimed water is called NEWater and is bottled directly from an advanced water purification facility for educational and celebratory purposes. Though most of the reused water is used for high-tech industry in Singapore, a small amount is returned to reservoirs for drinking water.

A 2012 study conducted by the National Research Council in the United States of America found that the risk of exposure to certain microbial and chemical contaminants from drinking reclaimed water does not appear to be any higher than the risk experienced in at least some current drinking water treatment systems, and may be orders of magnitude lower. This report recommends adjustments to the federal regulatory framework that could enhance public health protection for both planned and unplanned (or de facto) reuse and increase public confidence in water reuse.

Modern technologies such as reverse osmosis and ultraviolet disinfection are commonly used when reclaimed water will be mixed with the drinking water supply. An experiment by the University of New South Wales reportedly showed a reverse osmosis system removed ethinylestradiol and paracetamol from the wastewater, even at 1000 times the expected concentration.

Aboard the International Space Station, astronauts have been able to drink recycled urine due to the introduction of the ECLSS system. The system costs $250 million and has been working since May 2009. The system recycles wastewater and urine back into potable water used for drinking, food preparation, and oxygen generation. This cuts back on the need for resupplying the space station so often.

Indirect Potable Reuse

Some municipalities are using and others are investigating Indirect Potable Reuse (IPR) of reclaimed water. For example, reclaimed water may be pumped into (subsurface recharge) or percolated down to (surface recharge) groundwater aquifers, pumped out, treated again, and finally used as drinking water. This technique may also be referred to as *groundwater recharging*. This includes slow processes of further multiple purification steps via the layers of earth/sand (absorption) and microflora in the soil (biodegradation).

Direct Potable Reuse

In a Direct Potable Reuse (DPR) scheme, water is put directly into pipelines that go to a water treatment plant or distribution system. Direct potable reuse may occur with or without "engineered storage" such as underground or above ground tanks. Communities in Texas have implemented DPR projects, and the state of California is studying the feasibility of developing DPR regulations.

Unplanned Potable reuse

Water reuse occurs in various ways throughout the world. It happens daily on rivers and other water bodies everywhere. If you live in a community downstream of another, chances are you are reusing its water and likewise communities downstream of you are most likely reusing your water. Unplanned Indirect Potable Use has existed even before the introduction of reclaimed water. Many cities already use water from rivers that contain effluent discharged from upstream sewage treatment plants. There are many large towns on the River Thames upstream of London (Oxford, Reading, Swindon, Bracknell) that discharge their treated sewage ("non-potable water") into the river, which is used to supply London with water downstream.

This phenomenon is also observed in the United States, where the Mississippi River serves as both the destination of sewage treatment plant effluent and the source of potable water. Research conducted in the 1960s by the London Metropolitan Water Board demonstrated that the maximum extent of recycling water is about 11 times before the taste of water induces nausea in sensitive individuals. This is caused by the buildup of inorganic ions such as Cl^-, SO_4^{2-}, K^+ and Na^+, which are not removed by conventional sewage treatment.

Space Travel

Wastewater reclamation can be especially important in relation to human spaceflight.

- In 1998, NASA announced it had built a human waste reclamation bioreactor designed for use in the International Space Station and a manned Mars mission. Human urine and feces are input into one end of the reactor and pure oxygen, pure water, and compost (huma-nure) are output from the other end. The soil could be used for growing vegetables, and the bioreactor also produces electricity.

Design Considerations

Distribution

Nonpotable reclaimed water is often distributed with a dual piping network that keeps reclaimed water pipes completely separate from potable water pipes. In the United States and some other countries, nonpotable reclaimed water is distributed in lavender (light purple) pipes to distinguish it from potable water. The use of the color purple for pipes carrying recycled water was pioneered by the Irvine Ranch Water District in Irvine, California.

In many cities using reclaimed water, it is now in such demand that consumers are only allowed to use it on assigned days. Some cities that previously offered unlimited reclaimed water at a flat rate are now beginning to charge citizens by the amount they use.

Reclamation Processes

Wastewater must pass through numerous systems before being returned to the environment. Here is a partial listing from one particular plant system:

- Barscreens - Barscreens remove large solids that are sent into a grinder. All solids are then dumped into a sewer pipe at a Treatment Plant.

- Primary Settling Tanks - Readily settable and floatable solids are removed from the waste-water. These solids are skimmed from the top and bottom of the tanks and sent to the Treatment Plant where it'll be turned into fertilizer.

- Biological Treatment - The wastewater is cleaned through a biological treatment meth-od that uses microorganisms, bacteria which digest the sludge and reduce the nutrient content. Air bubbles up to keep the organisms suspended and to supply oxygen to the aerobic bacteria so they can metabolize the food, convert it to energy, CO_2, and water, and reproduce more microorganisms. This helps to remove ammonia also through ni-trification.

- Secondary Settling Tanks - The force of the flow slows down as sewage enters these tanks, allowing the microorganisms to settle to the bottom. As they settle, other small particles suspended in the water are picked up, leaving behind clear wastewater. Some of the micro-organisms that settle to the bottom are returned to the system to be used again.

- Tertiary Treatment - Deep-bed, single-media, gravity sand filters receive water from the secondary basins and filter out the remaining solids. As this is the final process to remove solids, the water in these filters is almost completely clear.

- Chlorine Contact Tanks - Three chlorine contact tanks disinfect the water to decrease the risks associated with discharging wastewater containing human pathogens. This step pro-tects the quality of the waters that receive the wastewater discharge.

One of two procedures are then followed according to the future disposal site:

1. Reclaimed Water Pump Station - The pump station distributes reclaimed water to users around the City. This may include golf courses, agricultural uses, cooling towers, or in land fills.

2. Water is passed through high level purification to be returned to the environment. Currently this means a reverse osmosis system.

Treatment Improvements

As world populations require both more clean water and better ways to dispose of wastewater, the demand for water reclamation will increase. Future success in water reuse will depend on whether this can be done without adverse effects on human health and the environment.

In the United States, reclaimed waste water is generally treated to secondary level when used for irrigation, but there are questions about the adequacy of that treatment. Some leading scientists in the main water society, AWWA, have long believed that secondary treatment is insufficient to protect people against pathogens, and recommend adding at least membrane filtration, reverse osmosis, ozonation, or other advanced treatments for irrigation water.

There have been recent advances in reverse osmosis in different countries, but have consistently produced very high quality water all the same. In Singapore, reclaimed water, also known as NE-Water has become cleaner than the government issue tap water. Also, according to Bartels, the Bedok Demonstration Plant, which uses RO membranes, has successfully run for the past 3 years, producing high quality wastewater all the while.

Seepage of nitrogen and phosphorus into ground and surface water is also becoming a serious problem, and will probably lead to at least tertiary treatment of reclaimed water to remove nutrients in the future. Even using secondary treatment, water quality can be improved. Water quality can also be improved as it passes through the subsurface mixing zone where surface water and groundwater combine. Testing for pathogens using Polymerase Chain Reaction (PCR) instead of older culturing techniques, and changing the discredited fecal coloform "indicator organism" standard would be improvements.

In a large study treatment plants showed that they could significantly reduce the numbers of parasites in effluent, just by making adjustments to the currently used process. But, even using the best of current technology, risk of spreading drug resistance in the environment through wastewater effluent, would remain.

Some scientists have suggested that there need to be basic changes in treatment, such as using bacteria to degrade waste based on nitrogen (urine) and not just carbonaceous (fecal) waste, saying that this would greatly improve effectiveness of treatment. Currently designed plants do not deal well with contaminants in solution (e.g. pharmaceuticals). "Dewatering" solids is a major problem. Some wastes could be disposed of without mixing them with water to begin with. In an interesting innovation, solids (sludge) could be removed before entering digesters and burned into a gas that could be used to run engines.

Emerging disinfection technologies include ultrasound, pulse arc electrohydraulic discharge, and bank filtration. Another issue is concern about weakened mandates for pretreatment of industrial wastes before they are made part of the municipal waste stream. Some also believe that hospitals should treat their own wastes. The safety of drinking reclaimed water which has been given advanced treatment and blended with other waters, remains controversial.

In recent years, as hydraulic fracturing of oil and gas formations has become more and more commonplace, new technologies for water recycling have emerged. One such technology uses a combination of ozone and electrocoagulation. This process removes organics, hydrocarbons, spent polymers, chemical additives used in the fracturing process, and heavy metals such as barium, iron, boron and more.

Alternatives

Seawater Desalination

In urban areas where climate change has threatened long-term water security and reduced rainfall over catchment areas, using reclaimed water for indirect potable use may be superior to other water supply augmentation methods. One other commonly used option is seawater desalination. Recycling wastewater and desalinating seawater may have many of the same disadvantages, including high costs of water treatment, infrastructure construction, transportation, and waste disposal problems. Although the best option varies from region to region, desalination is often superior economically, as reclaimed water usually requires a dual piping network, often with additional storage tanks, when used for nonpotable use.

Greywater Systems

A less elaborate alternative to reclaimed water is a greywater system. Greywater is wastewater that has been used in sinks, baths, showers, or washing machines, but does not contain sewage and has not been treated at the same levels as recycled water. In a home system, treated or untreated greywater may be used to flush toilets or for irrigation. Some systems now exist which directly use greywater from a sink to flush a toilet.

Rainwater Harvesting

Perhaps the simplest option is a rainwater harvesting system. Although there are concerns about the quality of rainwater in urban areas, due to air pollution and acid rain, many systems exist now to use untreated rainwater for nonpotable uses or treated rainwater for direct potable use. Urban design systems which incorporate rainwater harvesting and reduce runoff are known as Water Sensitive Urban Design (WSUD) in Australia, Low Impact Development (LID) in the United States and Sustainable urban drainage systems (SUDS) in the United Kingdom. There are also concerns about rainwater harvesting systems reducing the amount of run-off entering natural bodies of water.

Health Aspects

Reclaimed water is highly engineered for safety and reliability so that the quality of reclaimed water is more predictable than many existing surface and groundwater sources. Reclaimed water is considered safe when appropriately used. Reclaimed water planned for use in recharging aquifers or augmenting surface water receives adequate and reliable treatment before mixing with naturally occurring water and undergoing natural restoration processes. Some of this water eventually becomes part of drinking water supplies.

A water quality study published in 2009 compared the water quality differences of reclaimed/recycled water, surface water, and groundwater. Results indicate that reclaimed water, surface water, and groundwater are more similar than dissimilar with regard to constituents. The researchers tested for 244 representative constituents typically found in water. When detected, most constituents were in the parts per billion and parts per trillion range. DEET (a bug repellant), and Caffeine were found in all water types and virtually in all samples. Triclosan (in anti-bacterial soap & toothpaste) was found in all water types, but detected in higher levels (parts per trillion) in reclaimed water than in surface or groundwater. Very few hormones/steroids were detected in samples, and when detected were at very low levels. Haloacetic acids (a disinfection by-product) were found in all types of samples, even groundwater. The largest difference between reclaimed water and the other waters appears to be that reclaimed water has been disinfected and thus has disinfection by-products (due to chlorine use).

A 2005 study titled "Irrigation of Parks, Playgrounds, and Schoolyards with Reclaimed Water" found that there had been no incidences of illness or disease from either microbial pathogens or chemicals, and the risks of using reclaimed water for irrigation are not measurably different from irrigation using potable water. Studies by the National Academies of Science, the Monterey Regional Water Pollution Control Agency, and others have found reclaimed water to be safe for agricultural use.

Testing Standards

Reclaimed water is not regulated by the Environmental Protection Agency (EPA), but the EPA has developed water reuse guidelines that were most recently updated in 2012. The EPA Guidelines for Water Reuse represents the international standard for best practices in water reuse. The document was developed under a Cooperative Research and Development Agreement between the U.S. Environmental Protection Agency (EPA), the U.S. Agency for International Development (USAID), and the global consultancy CDM Smith. The Guidelines provide a framework for states to develop regulations that incorporate the best practices and address local requirements.

Ongoing wastewater research sometimes raise concerns about pathogens in the water. Many pathogens cannot be detected by currently used tests.

Recent literature also questions the validity of testing for "indicator organisms" instead of pathogens. Nor do present standards consider interactions of heavy metals and pharmaceuticals which may foster the development of drug resistant pathogens in waters derived from sewage.

To address these concerns about the source water, reclaimed water providers use multi-barrier treatment processes and constant monitoring to ensure that reclaimed water is safe and treated properly for the intended end use.

Potable Use

The main health risk for potable use of reclaimed water is the potential for pharmaceutical and other household chemicals or their derivatives (Environmental persistent pharmaceutical pollutants) to persist in this water. This would be of much less concern if the population were to keep their excrement out of the wastewater e.g. via the use of the Urine-diverting dry toilet or systems that treat blackwater separately from greywater.

Environmental Aspects

There is debate about possible health and environmental effects. To address these concerns, A Risk Assessment Study of potential health risks of recycled water and comparisons to conventional Pharmaceuticals and Personal Care Product (PPCP) exposures was conducted by the WateReuse Research Foundation. For each of four scenarios in which people come into contact with recycled water used for irrigation - children on the playground, golfers, and landscape, and agricultural workers - the findings from the study indicate that it could take anywhere from a few years to millions of years of exposure to nonpotable recycled water to reach the same exposure to PPCPs that we get in a single day through routine activities.

Using reclaimed water for non-potable uses saves potable water for drinking, since less potable water will be used for non-potable uses.

It sometimes contains higher levels of nutrients such as nitrogen, phosphorus and oxygen which may somewhat help fertilize garden and agricultural plants when used for irrigation.

The usage of water reclamation decreases the pollution sent to sensitive environments. It can also enhance wetlands, which benefits the wildlife depending on that eco-system. It also helps to stop the chances of drought as recycling of water reduces the use of fresh water supply from underground sources. For instance, The San Jose/Santa Clara Water Pollution Control Plant instituted a water recycling program to protect the San Francisco Bay area's natural salt water marshes.

Costs and Evaluation

The cost of reclaimed water exceeds that of potable water in many regions of the world, where a fresh water supply is plentiful. However, reclaimed water is usually sold to citizens at a cheaper rate to encourage its use. As fresh water supplies become limited from distribution costs, increased population demands, or climate change reducing sources, the cost ratios will evolve also. The evaluation of reclaimed water needs to consider the entire water supply system, as it may bring important value of flexibility into the overall system

History

Storm and sanitary sewers were necessarily developed along with the growth of cities. By the 1840s the luxury of indoor plumbing, which mixes human waste with water and flushes it away, eliminated the need for cesspools. Odor was considered the big problem in waste disposal and to address it, sewage could be drained to a lagoon, or "settled" and the solids removed, to be disposed of separately. This process is now called "primary treatment" and the settled solids are called "sludge."

At the end of the 19th century, since primary treatment still left odor problems, it was discovered that bad odors could be prevented by introducing oxygen into the decomposing sewage. This was the beginning of the biological aerobic and anaerobic treatments which are fundamental to waste water processes.

By the 1920s, it became necessary to further control the pollution caused by the large quantities of human and industrial liquid wastes which were being piped into rivers and oceans, and modern treatment plants were being built in the US and other industrialized nations by the 1930s.

Designed to make water safe for fishing and recreation, the Clean Water Act of 1972 mandated elimination of the discharge of untreated waste from municipal and industrial sources, and the US federal government provided billions of dollars in grants for building sewage treatment plants around the country. Modern treatment plants, usually using oxidation and/or chlorination in addition to primary and secondary treatment, were required to meet certain standards.

Current treatment improves the quality of separated wastewater solids or sludge. The separated water is given further treatment considered adequate for non potable use by local agencies, and discharged into bodies of water, or reused as reclaimed water. In places like Florida, where it is necessary to avoid nutrient overload of sensitive receiving water, reuse of treated or reclaimed water can be more economically feasible than meeting the higher standards for surface water disposal mandated by the Clean Water Act

Examples

Indirect Potable Reuse (IPR)

- Orange County, California

- Pasadena, California

- Singapore (where it is branded as *NEWater*)

- Payson, Arizona

- The Torreele project in the Veurne coastal region of Belgium, which began operating in 2002

- Virginia Occoquan Reservoir - The Upper Occoquan Sewage Authority plant discharges its highly treated output to supply roughly 20% of the inflow into the Occoquan Reservoir, which provides drinking water used by the Fairfax County Water Authority - one of the three major water providers in the Washington, D.C. metropolitan area.

Non-Potable Reuse (NPR)

- Austin, Texas

- Caboolture and Maroochy (South East Queensland, Australia) LGA's currently provide reclaimed water for industrial use (primarily capital works). Users must apply for a key to be able to access the compounds in which the outlets are located.

- Clark County, Nevada

- Clearwater, Florida

- Contra Costa County, California

- Melbourne, Australia

- Mount Buller Ski resort uses recycled water for snow making.

- San Antonio operates the largest recycled water system in the United States.

- Sydney, Australia

- Tucson, Arizona

- San Diego, California (San Diego County)

- St. Petersburg, Florida

Proposed

In some places, reclaimed water has been proposed for either potable or non-potable use:

- South East Queensland, Australia (planned for potable use as of late 2010)

- Newcastle, New South Wales, Australia (proposed for non-potable use).

- Canberra, Australian Capital Territory, Australia (proposed in January 2007 as a backup source of potable water)

- Los Angeles, California - By 2019, the Los Angeles Department of Water and Power will build a plant to replenish their groundwater aquifer with purified water in order to deal with the shortage of rain and snow fall, restricted water imports and local groundwater contamination.

Israel

As of 2010, Israel leads the world in the proportion of water it recycles. Israel treats 80% of its sewage (400 billion liters a year), and 100% of the sewage from the Tel Aviv metropolitan area is treated and reused as irrigation water for agriculture and public works. The remaining sludge is currently pumped into the Mediterranean, however a new bill has passed stating a conversion to treating the sludge to be used as manure. Only 20% of the treated water is lost (due to evaporation, leaks, overflows and seeping). The recycled water allows farmers to plan ahead and not be limited by water shortages. There are many levels of treatment, and many different ways of treating the water—which leads to a big difference in the quality of the end product. The best quality of reclaimed sewage water comes from adding a gravitational filtering step, after the chemical and biological cleansing. This method uses small ponds in which the water seeps through the sand into the aquifer in about 400 days, then is pumped out as clear purified water. This is nearly the same process used in the space station water recycling system, which turns urine and feces into purified drinking water, oxygen and manure.

To add to the efficiency of the Israeli system - the reclaimed sewage water may be mixed with reclaimed sea water (Plans are in action to increase the desalinization program up to 50% of the countries usage by 2013 - 600 billion liters of drinkable sea water a year), along with aquifer water and fresh sweet lake water - monitored by computer to account for the nationwide needs and input. This action reduced the outdated risk of salt and mineral percentages in the water. Plans to implement this overall usage of reclaimed water for drinking are discouraged by the psychological preconception of the public for the quality of reclaimed water, and the fear of its origin. As of today, all the reclaimed sewage water in Israel is used for agricultural and land improvement purposes.

U.S.

The leaders in use of reclaimed water in the U.S. are Florida and California, with Irvine Ranch Water District as one of the leading developers. They were the first district to approve the use of reclaimed water for in-building piping and use in flushing toilets.

In a January 2012 U.S. National Research Council report, a committee of independent experts found that expanding the reuse of municipal wastewater for irrigation, industrial uses, and drinking water augmentation could significantly increase the United States' total available water resources. The committee noted that a portfolio of treatment options is available to mitigate water quality issues in reclaimed water. The report also includes a risk analysis that suggests the risk of exposure to certain microbial and chemical contaminants from drinking reclaimed water is not any higher than the risk from drinking water from current water treatment systems—and in some cases, may be orders of magnitude lower. The report concludes that adjustments to the federal regulatory framework could enhance public health protection and increase public confidence in water reuse.

Australia

As Australia continues to battle the 7–10-year drought, nationwide, reclaimed effluent is becoming a popular option. Two major capital cities in Australia, Adelaide and Brisbane, have already committed to adding reclaimed effluent to their dwindling dams. The former has also built a desalination plant to help battle any future water shortages. Brisbane has been seen as a leader in this trend, and other cities and towns will review the Western Corridor Recycled Water Project once completed. Goulbourn, Canberra, Newcastle, and Regional Victoria, Australia are already considering building a reclaimed effluent process.

European Union

The second largest waste reclamation program in the world is in Spain, where 12% of the nation's waste is treated.

According to an EU-funded study, "Europe and the Mediterranean countries are lagging behind" California, Japan, and Australia "in the extent to which reuse is being taken up." According to the study "the concept (of reuse) is difficult for the regulators and wider public to understand and accept."

References

- Wastewater engineering : treatment and reuse (4th ed.). Metcalf & Eddy, Inc., McGraw Hill, USA. 2003. p. 1138. ISBN 0-07-112250-8.

- Tchobanoglous, G., Burton, F.L., and Stensel, H.D. (2003). Wastewater Engineering (Treatment Disposal Reuse) / Metcalf & Eddy, Inc. (4th ed.). McGraw-Hill Book Company. ISBN 0-07-041878-0.

- Khopkar, S. M. (2004). Environmental Pollution Monitoring And Control. New Delhi: New Age International. p. 299. ISBN 81-224-1507-5.

- Martin V. Melosi (2010). The Sanitary City: Environmental Services in Urban America from Colonial Times to the Present. University of Pittsburgh Press. p. 110. ISBN 9780822973379.

- Colin A. Russell (2003). Edward Frankland: Chemistry, Controversy and Conspiracy in Victorian England. Cambridge University Press. pp. 372–380. ISBN 9780521545815.

- Sharma, Sanjay Kumar; Sanghi, Rashmi (2012). Advances in Water Treatment and Pollution Prevention. Springer. ISBN 9789400742048. Retrieved 2013-02-07.

- Tilley, David F. (2011). Aerobic Wastewater Treatment Processes: History and Development. IWA Publishing. ISBN 9781843395423. Retrieved 2013-02-07.

- Water Reuse: Potential for Expanding the Nation's Water Supply through Reuse of Municipal Wastewater. National Research Council. 2012. ISBN 978-0-309-25749-7.

- Crook, James (2005). Irrigation of Parks, Playgrounds, and Schoolyards: Extent and Safety. Alexandria, VA: WateReuse Research Foundation. p. 60. ISBN 0-9747586-3-9.

- Ashton, John; Ubido, Janet (1991). "The Healthy City and the Ecological Idea" (PDF). Journal of the Society for the Social History of Medicine. 4 (1): 173–181. doi:10.1093/shm/4.1.173. Retrieved 8 July 2013.

- Environment Agency (archive) – Persistent, bioaccumulative and toxic PBT substances at the Wayback Machine (archived August 4, 2006). environment-agency.gov.uk. Retrieved on 2012-12-19.

- Natural Environmental Research Council – River sewage pollution found to be disrupting fish hormones. Planetearth.nerc.ac.uk. Retrieved on 2012-12-19.

- Ong, Hian Hai; Ryck, Luc De, "Best Sourcing Approach Keeps Water Production Costs Down", Water & Wastewater International: 13, retrieved 22 March 2012

- "Monterey Wastewater Reclamation Study for Agriculture, Final Report". Monterey Regional Water Pollution Control Agency. Retrieved 18 October 2011.

Water Purification and its Methods

The process of removing undesirable chemicals or biological contaminants from contaminated water is water purification. Most of the water purified is purified for human consumption, but there can be other purposes as well. The methods of water purification explained in the following script are filtration, sedimentation and water chlorination. The major components of water purification are discussed in this chapter.

Water Purification

Water purification is the process of removing undesirable chemicals, biological contaminants, suspended solids and gases from contaminated water. The goal is to produce water fit for a specific purpose. Most water is disinfected for human consumption (drinking water), but water purification may also be designed for a variety of other purposes, including fulfilling the requirements of medical, pharmacological, chemical and industrial applications. The methods used include physical processes such as filtration, sedimentation, and distillation; biological processes such as slow sand filters or biologically active carbon; chemical processes such as flocculation and chlorination and the use of electromagnetic radiation such as ultraviolet light.

Control room and schematics of the water purification plant to Lac de Bret, Switzerland

Purifying water may reduce the concentration of particulate matter including suspended particles, parasites, bacteria, algae, viruses, fungi, as well as reducing the amount of a range of dissolved and particulate material derived from the surfaces that come from runoff due to rain.

The standards for drinking water quality are typically set by governments or by international standards. These standards usually include minimum and maximum concentrations of contaminants, depending on the intended purpose of water use.

Visual inspection cannot determine if water is of appropriate quality. Simple procedures such as boiling or the use of a household activated carbon filter are not sufficient for treating all the possible contaminants that may be present in water from an unknown source. Even natural spring water – considered safe for all practical purposes in the 19th century – must now be tested before determining what kind of treatment, if any, is needed. Chemical and microbiological analysis, while expensive, are the only way to obtain the information necessary for deciding on the appropriate method of purification.

According to a 2007 World Health Organization (WHO) report, 1.1 billion people lack access to an improved drinking water supply, 88 percent of the 4 billion annual cases of diarrheal disease are attributed to unsafe water and inadequate sanitation and hygiene, while 1.8 million people die from diarrheal diseases each year. The WHO estimates that 94 percent of these diarrheal cases are preventable through modifications to the environment, including access to safe water. Simple techniques for treating water at home, such as chlorination, filters, and solar disinfection, and storing it in safe containers could save a huge number of lives each year. Reducing deaths from waterborne diseases is a major public health goal in developing countries.

Sources of Water

1. Groundwater: The water emerging from some deep ground water may have fallen as rain many tens, hundreds, or thousands of years ago. Soil and rock layers naturally filter the ground water to a high degree of clarity and often, it does not require additional treatment besides adding chlorine or chloramines as secondary disinfectants. Such water may emerge as springs, artesian springs, or may be extracted from boreholes or wells. Deep ground water is generally of very high bacteriological quality (i.e., pathogenic bacteria or the pathogenic protozoa are typically absent), but the water may be rich in dissolved solids, especially carbonates and sulfates of calcium and magnesium. Depending on the strata through which the water has flowed, other ions may also be present including chloride, and bicarbonate. There may be a requirement to reduce the iron or manganese content of this water to make it acceptable for drinking, cooking, and laundry use. Primary disinfection may also be required. Where groundwater recharge is practised (a process in which river water is injected into an aquifer to store the water in times of plenty so that it is available in times of drought), the groundwater may require additional treatment depending on applicable state and federal regulations.

2. Upland lakes and reservoirs: Typically located in the headwaters of river systems, upland reservoirs are usually sited above any human habitation and may be surrounded by a protective zone to restrict the opportunities for contamination. Bacteria and pathogen levels are usually low, but some bacteria, protozoa or algae will be present. Where uplands are forested or peaty, humic acids can colour the water. Many upland sources have low pH which require adjustment.

3. Rivers, canals and low land reservoirs: Low land surface waters will have a significant bacterial load and may also contain algae, suspended solids and a variety of dissolved constituents.

4. Atmospheric water generation is a new technology that can provide high quality drinking water by extracting water from the air by cooling the air and thus condensing water vapor.

5. Rainwater harvesting or fog collection which collect water from the atmosphere can be used especially in areas with significant dry seasons and in areas which experience fog even when there is little rain.

6. Desalination of seawater by distillation or reverse osmosis.

7. Surface Water: Freshwater bodies that are open to the atmosphere and are not designated as groundwater are termed surface waters.

Treatment

Aims

The aims of the treatment are to remove unwanted constituents in the water and to make it safe to drink or fit for a specific purpose in industry or medical applications. Widely varied techniques are available to remove contaminants like fine solids, micro-organisms and some dissolved inorganic and organic materials, or environmental persistent pharmaceutical pollutants. The choice of method will depend on the quality of the water being treated, the cost of the treatment process and the quality standards expected of the processed water.

The processes below are the ones commonly used in water purification plants. Some or most may not be used depending on the scale of the plant and quality of the raw (source) water.

Pre-treatment

1. Pumping and containment – The majority of water must be pumped from its source or directed into pipes or holding tanks. To avoid adding contaminants to the water, this physical infrastructure must be made from appropriate materials and constructed so that accidental contamination does not occur.

2. Screening – The first step in purifying surface water is to remove large debris such as sticks, leaves, rubbish and other large particles which may interfere with subsequent purification steps. Most deep groundwater does not need screening before other purification steps.

3. Storage – Water from rivers may also be stored in bankside reservoirs for periods between a few days and many months to allow natural biological purification to take place. This is especially important if treatment is by slow sand filters. Storage reservoirs also provide a buffer against short periods of drought or to allow water supply to be maintained during transitory pollution incidents in the source river.

4. Pre-chlorination – In many plants the incoming water was chlorinated to minimize the growth of fouling organisms on the pipe-work and tanks. Because of the potential adverse quality effects, this has largely been discontinued.

pH Adjustment

Pure water has a pH close to 7 (neither alkaline nor acidic). Sea water can have pH values that range from 7.5 to 8.4 (moderately alkaline). Fresh water can have widely ranging pH values depending on the geology of the drainage basin or aquifer and the influence of contaminant inputs (acid rain). If the water is acidic (lower than 7), lime, soda ash, or sodium hydroxide can be added to raise the pH during water purification processes. Lime addition increases the calcium ion concentration, thus raising the water hardness. For highly acidic waters, forced draft degasifiers can be an effective way to raise the pH, by stripping dissolved carbon dioxide from the water. Making the water alkaline helps coagulation and flocculation processes work effectively and also helps to minimize the risk of lead being dissolved from lead pipes and from lead solder in pipe fittings. Sufficient alkalinity also reduces the corrosiveness of water to iron pipes. Acid (carbonic acid, hydrochloric acid or sulfuric acid) may be added to alkaline waters in some circumstances to lower the pH. Alkaline water (above pH 7.0) does not necessarily mean that lead or copper from the plumbing system will not be dissolved into the water. The ability of water to precipitate calcium carbonate to protect metal surfaces and reduce the likelihood of toxic metals being dissolved in water is a function of pH, mineral content, temperature, alkalinity and calcium concentration.

Coagulation and Flocculation

One of the first steps in a conventional water purification process is the addition of chemicals to assist in the removal of particles suspended in water. Particles can be inorganic such as clay and silt or organic such as algae, bacteria, viruses, protozoa and natural organic matter. Inorganic and organic particles contribute to the turbidity and color of water.

The addition of inorganic coagulants such as aluminum sulfate (or alum) or iron (III) salts such as iron(III) chloride cause several simultaneous chemical and physical interactions on and among the particles. Within seconds, negative charges on the particles are neutralized by inorganic coagulants. Also within seconds, metal hydroxide precipitates of the aluminum and iron (III) ions begin to form. These precipitates combine into larger particles under natural processes such as Brownian motion and through induced mixing which is sometimes referred to as flocculation. The term most often used for the amorphous metal hydroxides is "floc." Large, amorphous aluminum and iron (III) hydroxides adsorb and enmesh particles in suspension and facilitate the removal of particles by subsequent processes of sedimentation and filtration.

Aluminum hydroxides are formed within a fairly narrow pH range, typically: 5.5 to about 7.7. Iron (III) hydroxides can form over a larger pH range including pH levels lower than are effective for alum, typically: 5.0 to 8.5.

In the literature, there is much debate and confusion over the usage of the terms coagulation and flocculation—where does coagulation end and flocculation begin? In water purification plants, there is usually a high energy, rapid mix unit process (detention time in seconds) where the coagulant chemicals are added followed by flocculation basins (detention times range from 15 to 45 minutes) where low energy inputs turn large paddles or other gentle mixing devices to enhance the formation of floc. In fact, coagulation and flocculation processes are ongoing once the metal salt coagulants are added.

Organic polymers were developed in the 1960s as aids to coagulants and, in some cases, as replacements for the inorganic metal salt coagulants. Synthetic organic polymers are high molecular weight compounds that carry negative, positive or neutral charges. When organic polymers are added to water with particulates, the high molecular weight compounds adsorb onto particle surfaces and through interparticle bridging coalesce with other particles to form floc. PolyDADMAC is a popular cationic (positively charged) organic polymer used in water purification plants.

Sedimentation

Waters exiting the flocculation basin may enter the sedimentation basin, also called a clarifier or settling basin. It is a large tank with low water velocities, allowing floc to settle to the bottom. The sedimentation basin is best located close to the flocculation basin so the transit between the two processes does not permit settlement or floc break up. Sedimentation basins may be rectangular, where water flows from end to end, or circular where flow is from the centre outward. Sedimentation basin outflow is typically over a weir so only a thin top layer of water—that furthest from the sludge—exits.

In 1904, Allen Hazen showed that the efficiency of a sedimentation process was a function of the particle settling velocity, the flow through the tank and the surface area of tank. Sedimentation tanks are typically designed within a range of overflow rates of 0.5 to 1.0 gallons per minute per square foot (or 1.25 to 2.5 meters per hour). In general, sedimentation basin efficiency is not a function of detention time or depth of the basin. Although, basin depth must be sufficient so that water currents do not disturb the sludge and settled particle interactions are promoted. As particle concentrations in the settled water increase near the sludge surface on the bottom of the tank, settling velocities can increase due to collisions and agglomeration of particles. Typical detention times for sedimentation vary from 1.5 to 4 hours and basin depths vary from 10 to 15 feet (3 to 4.5 meters).

Inclined flat plates or tubes can be added to traditional sedimentation basins to improve particle removal performance. Inclined plates and tubes drastically increase the surface area available for particles to be removed in concert with Hazen's original theory. The amount of ground surface area occupied by a sedimentation basin with inclined plates or tubes can be far smaller than a conventional sedimentation basin.

Sludge Storage and Removal

As particles settle to the bottom of a sedimentation basin, a layer of sludge is formed on the floor of the tank which must be removed and treated. The amount of sludge generated is significant, often 3 to 5 percent of the total volume of water to be treated. The cost of treating and disposing of the sludge can impact the operating cost of a water treatment plant. The sedimentation basin may be equipped with mechanical cleaning devices that continually clean its bottom, or the basin can be periodically taken out of service and cleaned manually.

Floc Blanket Clarifiers

A subcategory of sedimentation is the removal of particulates by entrapment in a layer of suspended floc as the water is forced upward. The major advantage of floc blanket clarifiers is that they occupy a smaller footprint than conventional sedimentation. Disadvantages are that particle removal efficiency can be highly variable depending on changes in influent water quality and influent water flow rate.

Dissolved Air Flotation

When particles to be removed do not settle out of solution easily, dissolved air flotation (DAF) is often used. Water supplies that are particularly vulnerable to unicellular algae blooms and supplies with low turbidity and high colour often employ DAF. After coagulation and flocculation processes, water flows to DAF tanks where air diffusers on the tank bottom create fine bubbles that attach to floc resulting in a floating mass of concentrated floc. The floating floc blanket is removed from the surface and clarified water is withdrawn from the bottom of the DAF tank.

Filtration

After separating most floc, the water is filtered as the final step to remove remaining suspended particles and unsettled floc.

Rapid Sand Filters

The most common type of filter is a rapid sand filter. Water moves vertically through sand which often has a layer of activated carbon or anthracite coal above the sand. The top layer removes organic compounds, which contribute to taste and odour. The space between sand particles is larger than the smallest suspended particles, so simple filtration is not enough. Most particles pass through surface layers but are trapped in pore spaces or adhere to sand particles. Effective filtration extends into the depth of the filter. This property of the filter is key to its operation: if the top layer of sand were to block all the particles, the filter would quickly clog.

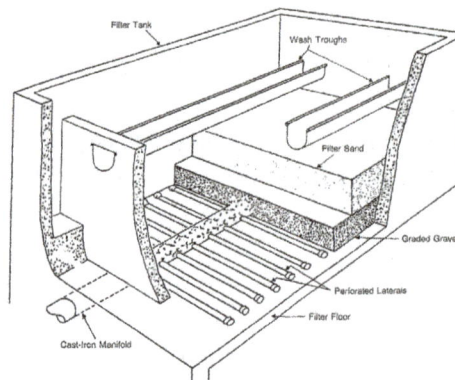

Cutaway view of a typical rapid sand filter

To clean the filter, water is passed quickly upward through the filter, opposite the normal direction (called *backflushing* or *backwashing*) to remove embedded particles. Prior to this step, compressed air may be blown up through the bottom of the filter to break up the compacted filter media to aid the backwashing process; this is known as *air scouring*. This contaminated water can be disposed of, along with the sludge from the sedimentation basin, or it can be recycled by mixing with the raw water entering the plant although this is often considered poor practice since it re-introduces an elevated concentration of bacteria into the raw water.

Some water treatment plants employ pressure filters. These work on the same principle as rapid gravity filters, differing in that the filter medium is enclosed in a steel vessel and the water is forced through it under pressure.

Advantages:

- Filters out much smaller particles than paper and sand filters can.

- Filters out virtually all particles larger than their specified pore sizes.

- They are quite thin and so liquids flow through them fairly rapidly.

- They are reasonably strong and so can withstand pressure differences across them of typically 2–5 atmospheres.

- They can be cleaned (back flushed) and reused.

Slow Sand Filters

Slow "artificial" filtration (a variation of bank filtration) to the ground, Water purification plant Káraný, Czech Republic

A profile of layers of gravel, sand and fine sand used in a slow sand filter plant.

Slow sand filters may be used where there is sufficient land and space, as the water must be passed very slowly through the filters. These filters rely on biological treatment processes for their action rather than physical filtration. The filters are carefully constructed using graded layers of sand, with the coarsest sand, along with some gravel, at the bottom and finest sand at the top. Drains at the base convey treated water away for disinfection. Filtration depends on the development of a thin biological layer, called the zoogleal layer or Schmutzdecke, on the surface of the filter. An effective slow sand filter may remain in service for many weeks or even months if the pre-treatment is well designed and produces water with a very low available nutrient level which physical methods of treatment rarely achieve. Very low nutrient levels allow water to be safely sent through distribution systems with very low disinfectant levels, thereby reducing consumer irritation over offensive levels of chlorine and chlorine by-products. Slow sand filters are not backwashed; they are maintained by having the top layer of sand scraped off when flow is eventually obstructed by biological growth.

A specific "large-scale" form of slow sand filter is the process of bank filtration, in which natural sediments in a riverbank are used to provide a first stage of contaminant filtration. While typically not clean enough to be used directly for drinking water, the water gained from the associated extraction wells is much less problematic than river water taken directly from the major streams where bank filtration is often used.

Membrane Filtration

Membrane filters are widely used for filtering both drinking water and sewage. For drinking water, membrane filters can remove virtually all particles larger than 0.2 μm—including *giardia* and *cryptosporidium*. Membrane filters are an effective form of tertiary treatment when it is desired to reuse the water for industry, for limited domestic purposes, or before discharging the water into a river that is used by towns further downstream. They are widely used in industry, particularly for beverage preparation (including bottled water). However no filtration can remove substances that are actually dissolved in the water such as phosphorus, nitrates and heavy metal ions.

Removal of Ions and other Dissolved Substances

Ultrafiltration membranes use polymer membranes with chemically formed microscopic pores that can be used to filter out dissolved substances avoiding the use of coagulants. The type of membrane media determines how much pressure is needed to drive the water through and what sizes of micro-organisms can be filtered out.

Ion exchange: Ion exchange systems use ion exchange resin- or zeolite-packed columns to replace unwanted ions. The most common case is water softening consisting of removal of Ca^{2+} and Mg^{2+} ions replacing them with benign (soap friendly) Na^+ or K^+ ions. Ion exchange resins are also used to remove toxic ions such as nitrite, lead, mercury, arsenic and many others.

Precipitative softening: Water rich in hardness (calcium and magnesium ions) is treated with lime (calcium oxide) and/or soda-ash (sodium carbonate) to precipitate calcium carbonate out of solution utilizing the common-ion effect.

Electrodeionization: Water is passed between a positive electrode and a negative electrode. Ion exchange membranes allow only positive ions to migrate from the treated water toward the negative electrode and only negative ions toward the positive electrode. High purity deionized water is produced continuously, similar to ion exchange treatment. Complete removal of ions from water is possible if the right conditions are met. The water is normally pre-treated with a reverse osmosis unit to remove non-ionic organic contaminants, and with gas transfer membranes to remove carbon dioxide. A water recovery of 99% is possible if the concentrate stream is fed to the RO inlet.

Disinfection

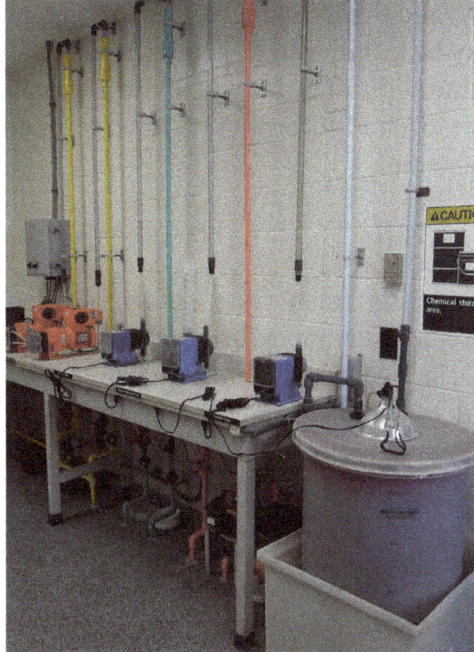

Pumps used to add required amount of chemicals to the clear water at the water purification plant before the distribution. From left to right: sodium hypochlorite for disinfection, zinc orthophosphate as a corrosion inhibitor, sodium hydroxide for pH adjustment, and fluoride for tooth decay prevention.

Disinfection is accomplished both by filtering out harmful micro-organisms and also by adding disinfectant chemicals. Water is disinfected to kill any pathogens which pass through the filters and to provide a residual dose of disinfectant to kill or inactivate potentially harmful micro-organisms in the storage and distribution systems. Possible pathogens include viruses, bacteria, including *Salmonella, Cholera, Campylobacter* and *Shigella*, and protozoa, including *Giardia lamblia* and other *cryptosporidia*. Following the introduction of any chemical disinfecting agent, the water is usually held in temporary storage – often called a contact tank or clear well to allow the disinfecting action to complete.

Chlorine Disinfection

The most common disinfection method involves some form of chlorine or its compounds such as chloramine or chlorine dioxide. Chlorine is a strong oxidant that rapidly kills many harmful micro-organisms. Because chlorine is a toxic gas, there is a danger of a release associated with its use. This problem is avoided by the use of sodium hypochlorite, which is a relatively inexpensive solution used in household bleach that releases free chlorine when dissolved in water. Chlorine solutions can be generated on site by electrolyzing common salt solutions. A solid form, calcium hypochlorite, releases chlorine on contact with water. Handling the solid, however, requires greater routine human contact through opening bags and pouring than the use of gas cylinders or bleach which are more easily automated. The generation of liquid sodium hypochlorite is both inexpensive and safer than the use of gas or solid chlorine.

All forms of chlorine are widely used, despite their respective drawbacks. One drawback is that chlorine from any source reacts with natural organic compounds in the water to form potentially harmful chemical by-products. These by-products, trihalomethanes (THMs) and haloacetic acids (HAAs), are both carcinogenic in large quantities and are regulated by the United States Environmental Protection Agency (EPA) and the Drinking Water Inspectorate in the UK. The formation of THMs and haloacetic acids may be minimized by effective removal of as many organics from the water as possible prior to chlorine addition. Although chlorine is effective in killing bacteria, it has limited effectiveness against protozoa that form cysts in water (*Giardia lamblia* and *Cryptosporidium*, both of which are pathogenic).

Chlorine Dioxide Disinfection

Chlorine dioxide is a faster-acting disinfectant than elemental chlorine. It is relatively rarely used, because in some circumstances it may create excessive amounts of chlorite, which is a by-product regulated to low allowable levels in the United States. Chlorine dioxide can be supplied as an aqueous solution and added to water to avoid gas handling problems; chlorine dioxide gas accumulations may spontaneously detonate.

Chloramine Disinfection

The use of chloramine is becoming more common as a disinfectant. Although chloramine is not as strong an oxidant, it does provide a longer-lasting residual than free chlorine and it will not readily form THMs or haloacetic acids. It is possible to convert chlorine to chloramine by adding ammonia to the water after addition of chlorine. The chlorine and ammonia react to form chloramine. Water distribution systems disinfected with chloramines may experience nitrification, as ammonia is a nutrient for bacterial growth, with nitrates being generated as a by-product.

Ozone Disinfection

Ozone is an unstable molecule which readily gives up one atom of oxygen providing a powerful oxidizing agent which is toxic to most waterborne organisms. It is a very strong, broad spectrum disinfectant that is widely used in Europe. It is an effective method to inactivate harmful protozoa that form cysts. It also works well against almost all other pathogens. Ozone is made by passing oxygen through ultraviolet light or a "cold" electrical discharge. To use ozone as a disinfectant, it must be created on-site and added to the water by bubble contact. Some of the advantages of ozone include the production of fewer dangerous by-products and the absence of taste and odour problems (in comparison to chlorination). Another advantage of ozone is that it leaves no residual disinfectant in the water. Ozone has been used in drinking water plants since 1906 where the first industrial ozonation plant was built in Nice, France. The U.S. Food and Drug Administration has accepted ozone as being safe; and it is applied as an anti-microbiological agent for the treatment, storage, and processing of foods. However, although fewer by-products are formed by ozonation, it has been discovered that ozone reacts with bromide ions in water to produce concentrations of the suspected carcinogen bromate. Bromide can be found in fresh water supplies in sufficient concentrations to produce (after ozonation) more than 10 parts per billion (ppb) of bromate — the maximum contaminant level established by the USEPA.

Ultraviolet Disinfection

Ultraviolet light (UV) is very effective at inactivating cysts, in low turbidity water. UV light's disinfection effectiveness decreases as turbidity increases, a result of the absorption, scattering, and shadowing caused by the suspended solids. The main disadvantage to the use of UV radiation is that, like ozone treatment, it leaves no residual disinfectant in the water; therefore, it is sometimes necessary to add a residual disinfectant after the primary disinfection process. This is often done through the addition of chloramines, discussed above as a primary disinfectant. When used in this manner, chloramines provide an effective residual disinfectant with very few of the negative effects of chlorination.

Portable Water Purification

Portable water purification devices and methods are available for disinfection and treatment in emergencies or in remote locations. Disinfection is the primary goal, since aesthetic considerations such as taste, odor, appearance, and trace chemical contamination do not affect the short-term safety of drinking water.

Additional Treatment Options

1. Water fluoridation: in many areas fluoride is added to water with the goal of preventing tooth decay. Fluoride is usually added after the disinfection process. In the U.S., fluoridation is usually accomplished by the addition of hexafluorosilicic acid, which decomposes in water, yielding fluoride ions.

2. Water conditioning: This is a method of reducing the effects of hard water. In water systems subject to heating hardness salts can be deposited as the decomposition of bicarbonate ions creates carbonate ions that precipitate out of solution. Water with high concentrations of hardness salts can be treated with soda ash (sodium carbonate) which precipitates out the excess salts, through the common-ion effect, producing calcium carbonate of very high purity. The precipitated calcium carbonate is traditionally sold to the manufacturers of toothpaste. Several other methods of industrial and residential water treatment are claimed (without general scientific acceptance) to include the use of magnetic and/or electrical fields reducing the effects of hard water.

3. Plumbosolvency reduction: In areas with naturally acidic waters of low conductivity (i.e. surface rainfall in upland mountains of igneous rocks), the water may be capable of dissolving lead from any lead pipes that it is carried in. The addition of small quantities of phosphate ion and increasing the pH slightly both assist in greatly reducing plumbosolvency by creating insoluble lead salts on the inner surfaces of the pipes.

4. Radium Removal: Some groundwater sources contain radium, a radioactive chemical element. Typical sources include many groundwater sources north of the Illinois River in Illinois. Radium can be removed by ion exchange, or by water conditioning. The back flush or sludge that is produced is, however, a low-level radioactive waste.

5. Fluoride Removal: Although fluoride is added to water in many areas, some areas of the world have excessive levels of natural fluoride in the source water. Excessive levels

can be toxic or cause undesirable cosmetic effects such as staining of teeth. Methods of reducing fluoride levels is through treatment with activated alumina and bone char filter media.

Other Water Purification Techniques

Other popular methods for purifying water, especially for local private supplies are listed below. In some countries some of these methods are also used for large scale municipal supplies. Particularly important are distillation (de-salination of seawater) and reverse osmosis.

1. Boiling: Bringing it to its boiling point at 100 °C (212 °F), is the oldest and most effective way since it eliminates most microbes causing intestine related diseases, but it cannot remove chemical toxins or impurities. For human health, complete sterilization of water is not required, since the heat resistant microbes are not intestine affecting. The traditional advice of boiling water for ten minutes is mainly for additional safety, since microbes start getting eliminated at temperatures greater than 60 °C (140 °F). Though the boiling point decreases with increasing altitude, it is not enough to affect the disinfecting process. In areas where the water is "hard" (that is, containing significant dissolved calcium salts), boiling decomposes the bicarbonate ions, resulting in partial precipitation as calcium carbonate. This is the "fur" that builds up on kettle elements, etc., in hard water areas. With the exception of calcium, boiling does not remove solutes of higher boiling point than water and in fact increases their concentration (due to some water being lost as vapour). Boiling does not leave a residual disinfectant in the water. Therefore, water that is boiled and then stored for any length of time may acquire new pathogens.

2. Granular Activated Carbon filtering: a form of activated carbon with a high surface area, adsorbs many compounds including many toxic compounds. Water passing through activated carbon is commonly used in municipal regions with organic contamination, taste or odors. Many household water filters and fish tanks use activated carbon filters to further purify the water. Household filters for drinking water sometimes contain silver as metallic silver nanoparticle. If water is held in the carbon block for longer period, microorganisms can grow inside which results in fouling and contamination. Silver nanoparticles are excellent anti-bacterial material and they can decompose toxic halo-organic compounds such as pesticides into non-toxic organic products.

3. Distillation involves boiling the water to produce water vapour. The vapour contacts a cool surface where it condenses as a liquid. Because the solutes are not normally vaporised, they remain in the boiling solution. Even distillation does not completely purify water, because of contaminants with similar boiling points and droplets of unvapourised liquid carried with the steam. However, 99.9% pure water can be obtained by distillation.

4. Reverse osmosis: Mechanical pressure is applied to an impure solution to force pure water through a semi-permeable membrane. Reverse osmosis is theoretically the most thorough method of large scale water purification available, although perfect semi-permeable membranes are difficult to create. Unless membranes are well-maintained, algae and other life forms can colonize the membranes.

5. The use of iron in removing arsenic from water.

6. Direct contact membrane distillation (DCMD). Applicable to desalination. Heated seawater is passed along the surface of a hydrophobic polymer membrane. Evaporated water passes from the hot side through pores in the membrane into a stream of cold pure water on the other side. The difference in vapour pressure between the hot and cold side helps to push water molecules through.

7. Desalination – is a process by which saline water (generally sea water) is converted to fresh water. The most common desalination processes are distillation and reverse osmosis. Desalination is currently expensive compared to most alternative sources of water, and only a very small fraction of total human use is satisfied by desalination. It is only economically practical for high-valued uses (such as household and industrial uses) in arid areas.

8. Gas hydrate crystals centrifuge method. If carbon dioxide or other low molecular weight gas is mixed with contaminated water at high pressure and low temperature, gas hydrate crystals will form exothermically. Separation of the crystalline hydrate may be performed by centrifuge or sedimentation and decanting. Water can be released from the hydrate crystals by heating

9. In Situ Chemical Oxidation, a form of advanced oxidation processes and advanced oxidation technology, is an environmental remediation technique used for soil and/or groundwater remediation to reduce the concentrations of targeted environmental contaminants to acceptable levels. ISCO is accomplished by injecting or otherwise introducing strong chemical oxidizers directly into the contaminated medium (soil or groundwater) to destroy chemical contaminants in place. It can be used to remediate a variety of organic compounds, including some that are resistant to natural degradation

Safety and Controversies

In April, 2007, the water supply of Spencer, Massachusetts became contaminated with excess sodium hydroxide (lye) when its treatment equipment malfunctioned.

Drinking water pollution detector Rainbow trout (*Oncorhynchus mykiss*) are being used in water purification plants to detect acute water pollution

Many municipalities have moved from free chlorine to chloramine as a disinfection agent. However, chloramine appears to be a corrosive agent in some water systems. Chloramine can dissolve the "protective" film inside older service lines, leading to the leaching of lead into residential spigots. This can result in harmful exposure, including elevated blood lead levels. Lead is a known neurotoxin.

Demineralized Water

Distillation removes all minerals from water, and the membrane methods of reverse osmosis and nanofiltration remove most to all minerals. This results in demineralized water which is not considered ideal drinking water. The World Health Organization has investigated the health effects of demineralized water since 1980. Experiments in humans found that demineralized water increased diuresis and the elimination of electrolytes, with decreased blood serum potassium concentration. Magnesium, calcium, and other minerals in water can help to protect against nutritional deficiency. Demineralized water may also increase the risk from toxic metals because it more readily leaches materials from piping like lead and cadmium, which is prevented by dissolved minerals such as calcium and magnesium. Low-mineral water has been implicated in specific cases of lead poisoning in infants, when lead from pipes leached at especially high rates into the water. Recommendations for magnesium have been put at a minimum of 10 mg/L with 20–30 mg/L optimum; for calcium a 20 mg/L minimum and a 40–80 mg/L optimum, and a total water hardness (adding magnesium and calcium) of 2 to 4 mmol/L. At water hardness above 5 mmol/L, higher incidence of gallstones, kidney stones, urinary stones, arthrosis, and arthropathies have been observed. Additionally, desalination processes can increase the risk of bacterial contamination.

Manufacturers of home water distillers claim the opposite—that minerals in water are the cause of many diseases, and that most beneficial minerals come from food, not water. They quote the American Medical Association as saying "The body's need for minerals is largely met through foods, not drinking water." The WHO report agrees that "drinking water, with some rare exceptions, is not the major source of essential elements for humans" and is "not the major source of our calcium and magnesium intake", yet states that demineralized water is harmful anyway. "Additional evidence comes from animal experiments and clinical observations in several countries. Animals given zinc or magnesium dosed in their drinking water had a significantly higher concentration of these elements in the serum than animals given the same elements in much higher amounts with food and provided with low-mineral water to drink."

History

Drawing of an apparatus for studying the chemical analysis of mineral waters in a book from 1799.

The first experiments into water filtration were made in the 17th century. Sir Francis Bacon attempted to desalinate sea water by passing the flow through a sand filter. Although his experiment did not succeed, it marked the beginning of a new interest in the field. The fathers of microscopy, Antonie van Leeuwenhoek and Robert Hooke, used the newly invented microscope to observe for the first time small material particles that lay suspended in the water, laying the groundwork for the future understanding of waterborne pathogens.

Sand Filter

The first documented use of sand filters to purify the water supply dates to 1804, when the owner of a bleachery in Paisley, Scotland, John Gibb, installed an experimental filter, selling his unwanted surplus to the public. This method was refined in the following two decades by engineers working for private water companies, and it culminated in the first treated public water supply in the world, installed by engineer James Simpson for the Chelsea Waterworks Company in London in 1829. This installation provided filtered water for every resident of the area, and the network design was widely copied throughout the United Kingdom in the ensuing decades.

Original map by John Snow showing the clusters of cholera cases in the London epidemic of 1854.

The practice of water treatment soon became mainstream and common, and the virtues of the system were made starkly apparent after the investigations of the physician John Snow during the 1854 Broad Street cholera outbreak. Snow was sceptical of the then-dominant miasma theory that stated that diseases were caused by noxious "bad airs". Although the germ theory of disease had not yet been developed, Snow's observations led him to discount the prevailing theory. His 1855 essay *On the Mode of Communication of Cholera* conclusively demonstrated the role of the water supply in spreading the cholera epidemic in Soho, with the use of a dot distribution map and statistical proof to illustrate the connection between the quality of the water source and cholera cases. His data convinced the local council to disable the water pump, which promptly ended the outbreak.

The Metropolis Water Act introduced the regulation of the water supply companies in London, including minimum standards of water quality for the first time. The Act "made provision for securing the supply to the Metropolis of pure and wholesome water", and required that all water be "effectually filtered" from 31 December 1855. This was followed up with legislation for the mandatory inspection of water quality, including comprehensive chemical analyses, in 1858. This legislation set a worldwide precedent for similar state public health interventions across Europe. The

Metropolitan Commission of Sewers was formed at the same time, water filtration was adopted throughout the country, and new water intakes on the Thames were established above Teddington Lock. Automatic pressure filters, where the water is forced under pressure through the filtration system, were innovated in 1899 in England.

Water Chlorination

John Snow was the first to successfully use chlorine to disinfect the water supply in Soho that had helped spread the cholera outbreak. William Soper also used chlorinated lime to treat the sewage produced by typhoid patients in 1879.

In a paper published in 1894, Moritz Traube formally proposed the addition of chloride of lime (calcium hypochlorite) to water to render it "germ-free." Two other investigators confirmed Traube's findings and published their papers in 1895. Early attempts at implementing water chlorination at a water treatment plant were made in 1893 in Hamburg, Germany and in 1897 the city of Maidstone England was the first to have its entire water supply treated with chlorine.

Permanent water chlorination began in 1905, when a faulty slow sand filter and a contaminated water supply led to a serious typhoid fever epidemic in Lincoln, England. Dr. Alexander Cruickshank Houston used chlorination of the water to stem the epidemic. His installation fed a concentrated solution of chloride of lime to the water being treated. The chlorination of the water supply helped stop the epidemic and as a precaution, the chlorination was continued until 1911 when a new water supply was instituted.

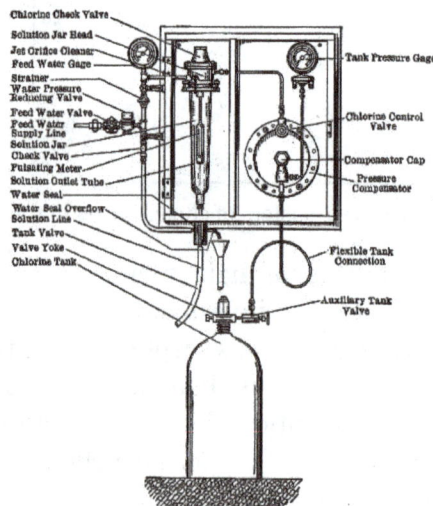

Manual-control chlorinator for the liquefaction of chlorine for water purification, early 20th century. From *Chlorination of Water* by Joseph Race, 1918.

The first continuous use of chlorine in the United States for disinfection took place in 1908 at Boonton Reservoir (on the Rockaway River), which served as the supply for Jersey City, New Jersey. Chlorination was achieved by controlled additions of dilute solutions of chloride of lime (calcium hypochlorite) at doses of 0.2 to 0.35 ppm. The treatment process was conceived by Dr. John L. Leal and the chlorination plant was designed by George Warren Fuller. Over the next few years, chlorine disinfection using chloride of lime were rapidly installed in drinking water systems around the world.

The technique of purification of drinking water by use of compressed liquefied chlorine gas was developed by a British officer in the Indian Medical Service, Vincent B. Nesfield, in 1903. According to his own account:

It occurred to me that chlorine gas might be found satisfactory ... if suitable means could be found for using it.... The next important question was how to render the gas portable. This might be accomplished in two ways: By liquefying it, and storing it in lead-lined iron vessels, having a jet with a very fine capillary canal, and fitted with a tap or a screw cap. The tap is turned on, and the cylinder placed in the amount of water required. The chlorine bubbles out, and in ten to fifteen minutes the water is absolutely safe. This method would be of use on a large scale, as for service water carts.

U.S. Army Major Carl Rogers Darnall, Professor of Chemistry at the Army Medical School, gave the first practical demonstration of this in 1910. Shortly thereafter, Major William J. L. Lyster of the Army Medical Department used a solution of calcium hypochlorite in a linen bag to treat water. For many decades, Lyster's method remained the standard for U.S. ground forces in the field and in camps, implemented in the form of the familiar Lyster Bag (also spelled Lister Bag). This work became the basis for present day systems of municipal water *purification.*

Filtration

Filtration is any of various mechanical, physical or biological operations that separate solids from fluids (liquids or gases) by adding a medium through which only the fluid can pass. The fluid that passes through is called the filtrate. In physical filters oversize solids in the fluid are retained and in biological filters particulates are trapped and ingested and metabolites are retained and removed. However, the separation is not complete; solids will be contaminated with some fluid and filtrate will contain fine particles (depending on the pore size ,filter thickness and biological activity). Filtration occurs both in nature and in engineered systems; there are biologic, geologic, and industrial forms. For example, in animals (including humans), renal filtration removes wastes from the blood, and in water treatment and sewage treatment, undesirable constituents are removed by absorption into a biological film grown on or in the filter medium, as in slow sand filtration.

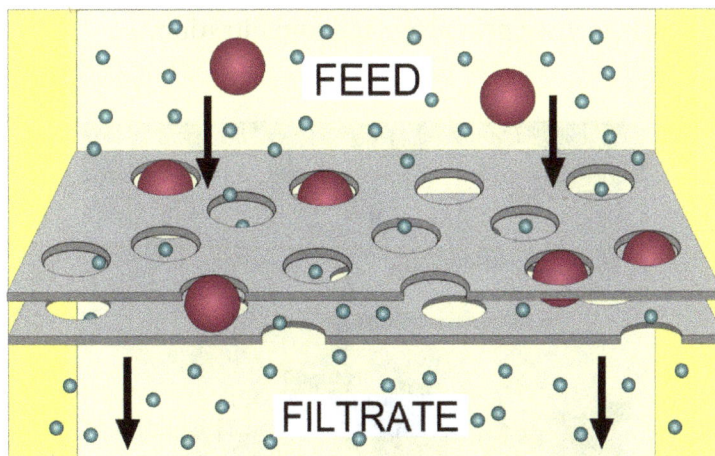

Diagram of simple filtration: oversize particles in the feed cannot pass through the lattice structure of the filter, while fluid and small particles pass through, becoming filtrate.

Applications

- Filtration is used to separate particles and fluid in a suspension, where the fluid can be a liquid, a gas or a supercritical fluid. Depending on the application, either one or both of the components may be isolated.

- Filtration, as a physical operation is very important in chemistry for the separation of materials of different chemical composition. A solvent is chosen which dissolves one component, while not dissolving the other. By dissolving the mixture in the chosen solvent, one component will go into the solution and pass through the filter, while the other will be retained. This is one of the most important techniques used by chemists to purify compounds.

- Filtration is also important and widely used as one of the unit operations of chemical engineering. It may be simultaneously combined with other unit operations to process the feed stream, as in the biofilter, which is a combined filter and biological digestion device.

- Filtration differs from sieving, where separation occurs at a single perforated layer (a sieve). In sieving, particles that are too big to pass through the holes of the sieve are retained. In filtration, a multilayer lattice retains those particles that are unable to follow the tortuous channels of the filter. Oversize particles may form a cake layer on top of the filter and may also block the filter lattice, preventing the fluid phase from crossing the filter (blinding). Commercially, the term filter is applied to membranes where the separation lattice is so thin that the surface becomes the main zone of particle separation, even though these products might be described as sieves.

- Filtration differs from adsorption, where it is not the physical size of particles that causes separation but the effects of surface charge. Some adsorption devices containing activated charcoal and ion exchange resin are commercially called filters, although filtration is not their principal function.

- Filtration differs from removal of magnetic contaminants from fluids with magnets (typically lubrication oil, coolants and fuel oils), because there is no filter medium. Commercial devices called "magnetic filters" are sold, but the name reflects their use, not their mode of operation.

The remainder of this article focuses primarily on liquid filtration.

Methods

Hot Filtration, solution contained in the Erlenmeyer flask is heated on a hot plate in order to prevent re-crystallization of solids in the flask itself

There are many different methods of filtration; all aim to attain the separation of substances. Separation is achieved by some form of interaction between the substance or objects to be removed and the filter. The substance that is to pass through the filter must be a fluid, i.e. a liquid or gas. Methods of filtration vary depending on the location of the targeted material, i.e. whether it is dissolved in the fluid phase or suspended as a solid.

There are several filtration techniques depending on the desired outcome namely, hot, cold and vacuum filtration. Some of the major purposes of getting a desired outcome are, for the removal of impurities from a mixture or, for the isolation of solids from a mixture.

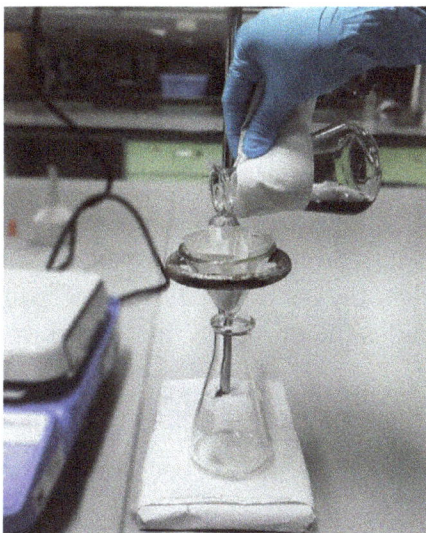

Hot Filtration for the separation of solids from a hot solution

Hot filtration method is mainly used to separate solids from a hot solution. This is done in order to prevent crystal formation in the filter funnel and other apparatuses that comes in contact with the solution. As a result, the apparatus and the solution used are heated in order to prevent the rapid decrease in temperature which in turn, would lead to the crystallization of the solids in the funnel and hinder the filtration process. One of the most important measure to prevent the formation of crystals in the funnel and to undergo effective hot filtration is the use stemless filter funnel. Due to the absence of stem in the filter funnel, there is a decrease in the surface area of contact between the solution and the stem of the filter funnel, hence preventing re-crystallization of solid in the funnel, adversely effecting filtration process.

Cold Filtration, the ice bath is used to cool down the temperature of the solution before undergoing the filtration process

Cold Filtration method is the use of ice bath in order to rapidly cool down the solution to be crystallized rather than leaving it out to cool it down slowly in the room temperature. This technique results to the formation of very small crystals as opposed to getting large crystals by cooling the solution down at room temperature.

Vacuum Filtration technique is most preferred for small batch of solution in order to quickly dry out small crystals. This method requires a Büchner funnel, filter paper of smaller diameter than the funnel, Büchner flask, and rubber tubing to connect to vacuum source.

Filter Media

Two main types of filter media are employed in any chemical laboratory— surface filter, a solid sieve which traps the solid particles, with or without the aid of filter paper (e.g. Büchner funnel, Belt filter, Rotary vacuum-drum filter, Cross-flow filters, Screen filter), and a depth filter, a bed of granular material which retains the solid particles as it passes (e.g. sand filter). The first type allows the solid particles, i.e. the residue, to be collected intact; the second type does not permit this. However, the second type is less prone to clogging due to the greater surface area where the particles can be trapped. Also, when the solid particles are very fine, it is often cheaper and easier to discard the contaminated granules than to clean the solid sieve.

Filter media can be cleaned by rinsing with solvents or detergents. Alternatively, in engineering applications, such as swimming pool water treatment plants, they may be cleaned by backwashing. Self-cleaning screen filters utilize point-of-suction backwashing to clean the screen without interrupting system flow.

Achieving Flow through the Filter

Fluids flow through a filter due to a difference in pressure — fluid flows from the high pressure side to the low pressure side of the filter, leaving some material behind. The simplest method to achieve this is by gravity and can be seen in the coffeemaker example. In the laboratory, pressure in the form of compressed air on the feed side (or vacuum on the filtrate side) may be applied to make the filtration process faster, though this may lead to clogging or the passage of fine particles. Alternatively, the liquid may flow through the filter by the force exerted by a pump, a method commonly used in industry when a reduced filtration time is important. In this case, the filter need not be mounted vertically.

Filter Aid

Certain filter aids may be used to aid filtration. These are often incompressible diatomaceous earth, or kieselguhr, which is composed primarily of silica. Also used are wood cellulose and other inert porous solids such as the cheaper and safer perlite.

These filter aids can be used in two different ways. They can be used as a precoat before the slurry is filtered. This will prevent gelatinous-type solids from plugging the filter medium and also give a clearer filtrate. They can also be added to the slurry before filtration. This increases the porosity of the cake and reduces resistance of the cake during filtration. In a rotary filter, the filter aid may be applied as a precoat; subsequently, thin slices of this layer are sliced off with the cake.

The use of filter aids is usually limited to cases where the cake is discarded or where the precipitate can be chemically separated from the filter.

Alternatives

Filtration is a more efficient method for the separation of mixtures than decantation, but is much more time consuming. If very small amounts of solution are involved, most of the solution may be soaked up by the filter medium.

An alternative to filtration is centrifugation — instead of filtering the mixture of solid and liquid particles, the mixture is centrifuged to force the (usually) denser solid to the bottom, where it often forms a firm cake. The liquid above can then be decanted. This method is especially useful for separating solids which do not filter well, such as gelatinous or fine particles. These solids can clog or pass through the filter, respectively.

Examples

Filter flask (suction flask, with sintered glass filter containing sample).
Note the almost colourless filtrate in the receiver flask.

Examples of filtration include

- The coffee filter to keep the coffee separate from the grounds.

- HEPA filters in air conditioning to remove particles from air.

- Belt filters to extract precious metals in mining.

- Horizontal plate filter, also known as Sparkler filter.

- Furnaces use filtration to prevent the furnace elements from fouling with particulates.

- Pneumatic conveying systems often employ filtration to stop or slow the flow of material that is transported, through the use of a baghouse.

- In the laboratory, a Büchner funnel is often used, with a filter paper serving as the porous barrier.

An experiment to prove the existence of microscopic organisms involves the comparison of water passed through unglazed porcelain and unfiltered water. When left in sealed containers the filtered water takes longer to go foul, demonstrating that very small items (such as bacteria) can be removed from fluids by filtration.

In the kidney, renal filtration is the filtration of blood in the glomerulus, followed by selective re-absorbtion of many substances essential for the body to maintain homeostasis.

Sedimentation (Water Treatment)

Sedimentation is a physical water treatment process using gravity to remove suspended solids from water. Solid particles entrained by the turbulence of moving water may be removed naturally by sedimentation in the still water of lakes and oceans. Settling basins are ponds constructed for the purpose of removing entrained solids by sedimentation. Clarifiers are tanks built with mechanical means for continuous removal of solids being deposited by sedimentation.

Basics

Suspended solids (or SS), is the mass of dry solids retained by a filter of a given porosity related to the volume of the water sample. This includes particles of a size not lower than 10 μm.

Colloids are particles of a size between 0.001 μm and 1 μm depending on the method of quantification. Because of Brownian motion and electrostatic forces balancing the gravity, they are not likely to settle naturally.

The limit sedimentation velocity of a particle is its theoretical descending speed in clear and still water. In settling process theory, a particle will settle only if:

1. In a vertical ascending flow, the ascending water velocity is lower than the limit sedimentation velocity.

2. In a longitudinal flow, the ratio of the length of the tank to the height of the tank is higher than the ratio of the water velocity to the limit sedimentation velocity.

Removal of suspended particles by sedimentation depends upon the size and specific gravity of those particles. Suspended solids retained on a filter may remain in suspension if their specific gravity is similar to water while very dense particles passing through the filter may settle. Settleable solids are measured as the visible volume accumulated at the bottom of an Imhoff cone after water has settled for one hour.

Gravitational theory is employed, alongside the derivation from Newton's second law and the Navier–Stokes equations.

$$V\_s = \sqrt{4/3 \, ((\rho\_p - \rho\_d)/\rho\_p)} \, (g d\_p)/C\_d \quad (1)$$

Stokes' law explains the relationship between the settling rate and the particle diameter. Under specific conditions, the particle settling rate is directly proportional to the square of particle diameter and inversely proportional to liquid viscosity.

The settling velocity, defined as the residence time taken for the particles to settle in the tank, enables the calculation of tank volume. Precise design and operation of a sedimentation tank is of high importance in order to keep the amount of sediment entering the diversion system to a minimum threshold by maintaining the transport system and stream stability to remove the sediment diverted from the system. This is achieved by reducing stream velocity as low as possible for the longest period of time possible. This is feasible by widening the approach channel and lowering its floor to reduce flow velocity thus allowing sediment to settle out of suspension due to gravity. The settling behavior of heavier particulates is also affected by the turbulence.

Designs

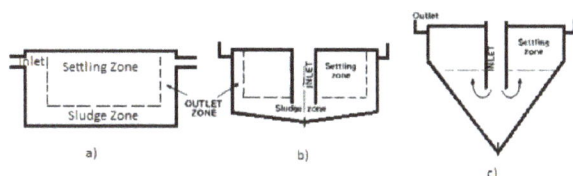

Different clarifier designs

Although sedimentation might occur in tanks of other shapes, removal of accumulated solids is easiest with conveyor belts in rectangular tanks or with scrapers rotating around the central axis of circular tanks. Settling basins and clarifiers should be designed based on the settling velocity of the smallest particle to be theoretically 100% removed. The overflow rate is defined as:

Overflow rate (V_o) = Flow of water (Q (cubic metre per second)) /(Surface area of settling basin (A))(m^2)

The unit of overflow rate is usually feet per second, a velocity. Any particle with settling velocity (Vs) greater than the overflow rate will settle out, while other particles will settle in the ratio Vs/Vo. There are recommendations on the overflow rates for each design that ideally take into account the change in particle size as the solids move through the operation:

- Quiescent zones: 0.031 ft/s

- Full-flow basins: 0.013 ft/s

- Off-line basins: 0.0015 ft/s

However, factors such as flow surges, wind shear, scour, and turbulence reduce the effectiveness of settling. To compensate for these less than ideal conditions, it is recommended doubling the area calculated by the previous equation. It is also important to equalize flow distribution at each point across the cross-section of the basin. Poor inlet and outlet designs can produce extremely poor flow characteristics for sedimentation.

Settling basins and clarifiers can be designed as long rectangles, that are hydraulically more stable and easier to control for large volumes. Circular clarifiers work as a common thickener (without the usage of rakes), or as upflow tanks.

Sedimentation efficiency does not depend on the tank depth. If the forward velocity is low enough so that the settled material does not re-suspend from the tank floor, the area is still the main parameter when designing a settling basin or clarifier, taking care that the depth is not too low.

Assessment of Main Process Characteristics

Settling basins and clarifiers are designed to retain water so that suspended solids can settle. By sedimentation principles, the suitable treatment technologies should be chosen depending on the specific gravity, size and shear resistance of particles. Depending on the size and density of particles, and physical properties of the solids, there are four types of sedimentation processes:

- Type 1 – Dilutes, non-flocculent, free-settling (every particle settles independently.)

- Type 2 – Dilute, flocculent (particles can flocculate as they settle).

- Type 3 – Concentrated suspensions, zone settling, hindered settling (sludge thickening).

- Type 4 – Concentrated suspensions, compression (sludge thickening).

Different factors control the sedimentation rate in each.

Settling of Discrete Particles

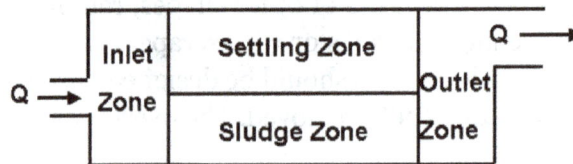

The four functional zones of a continuous flow settling basin

Unhindered settling is a process that removes the discrete particles in a very low concentration without interference from nearby particles. In general, if the concentration of the solutions is lower than 500 mg/L total suspended solids, sedimentation will be considered discrete. Concentrations of raceway effluent total suspended solids (TSS) in the west are usually less than 5 mg/L net. TSS concentrations of off-line settling basin effluent are less than 100 mg/L net. The particles keep their size and shape during discrete settling, with an independent velocity. With such low concentrations of suspended particles, the probability of particle collisions is very low and consequently the rate of floculation is small enough to be neglected for most calculations. Thus the surface area of the settling basin becomes the main factor of sedimentation rate. All continuous flow settling basins are divided into four parts: inlet zone, settling zone, sludge zone and outlet zone.

In the inlet zone, flow is established in a same forward direction. Sedimentation occurs in the settling zone as the water flow towards to outlet zone. The clarified liquid is then flow out from outlet zone. Sludge zone: settled will be collected here and usually we assume that it is removed from water flow once the particles arrives the sludge zone.

In an ideal rectangular sedimentation tank, in the settling zone, the critical particle enters at the top of the settling zone, and the settle velocity would be the smallest value to reach the sludge zone, and at the end of outlet zone, the velocity component of this critical particle are Vs, the settling velocity in vertical direction and Vh in horizontal direction.

From Figure 1, the time needed for the particle to settle;

$$t\_o = H/V\_p = L/Vs(3)$$

Since the surface area of the tank is WL, and Vs = Q/WL, Vh = Q/WH, where Q is the flow rate and W, L, H is the width, length, depth of the tank.

According to Eq. 1, this also is a basic factor that can control the sedimentation tank performance which called overflow rate.

Eq. 2 also tell us that the depth of sedimentation tank is independent to the sedimentation efficiency, only if the forward velocity is low enough to make sure the settled mass would not suspended again from the tank floor.

Settlement of Flocculent Particles

In a horizontal sedimentation tank, some particles may not follow the diagonal line in while settling faster as they grow. So this says that particles can grow and develop a higher settling velocity if a greater depth with longer retention time. However, the collision chance would be even greater if the same retention time were spread over a longer, shallower tank. In fact, in order to avoid hydraulic short-circuiting, tanks usually are made 3–6 m deep with retention times of a few hours.

Zone-settling Behaviour

As the concentration of particles in a suspension is increased, a point is reached where particles are so close together that they no longer settle independently of one another and the velocity fields of the fluid displaced by adjacent particles, overlap. There is also a net upward flow of liquid displaced by the settling particles. This results in a reduced particle-settling velocity and the effect is known as hindered settling.

There is a common case for hindered settling occurs. the whole suspension tends to settle as a 'blanket' due to its extremely high particle concentration. This is known as zone settling, because it is easy to make a distinction between several different zones which separated by concentration discontinuities. represents a typical batch-settling column tests on a suspension exhibiting zone-settling characteristics. There is a clear interface near the top of the column would be formed to separating the settling sludge mass from the clarified supernatant as long as leaving such a sus-pension to stand in a settling column. As the suspension settles, this interface will move down at the same speed. At the same time, there is an interface near the bottom between that settled sus-pension and the suspended blanket. After settling of suspension is complete, the bottom interface would move upwards and meet the top interface which moves downwards.

Compression Settling

The settling particles can contact each other and arise when approaching the floor of the sedimentation tanks at very high particle concentration. So that further settling will only occur in adjust matrix as the sedimentation rate decreasing. This is can be illustrated by the lower region of the zone-settling diagram. In Compression zone, the settled solids are compressed by gravity (the weight of solids), as the settled solids are compressed under the weight of overlying solids, and water is squeezed out while the space gets smaller.

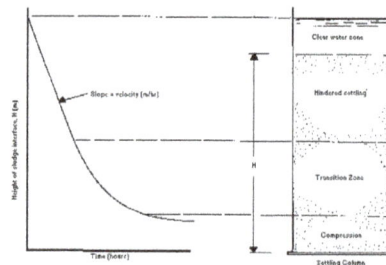

Typical batch-settling column test on a suspension exhibiting zone-settling characteristics

Applications

Potable Water Treatment

Sedimentation in potable water treatment generally follows a step of chemical coagulation and flocculation, which allows grouping particles together into flocs of a bigger size. This increases the settling speed of suspended solids and allows settling colloids.

Wastewater Treatment

Sedimentation has been used to treat wastewater for millennia.

Primary treatment of sewage is removal of floating and settleable solids through sedimentation. *Primary clariiers* reduce the content of suspended solids as well as the pollutant embedded in the suspended solids. Because of the large amount of reagent necessary to treat domestic wastewater, preliminary chemical coagulation and flocculation are generally not used, remaining suspended solids being reduced by following stages of the system. However, coagulation and flocculation can be used for building a compact treatment plant (also called a "package treatment plant"), or for further polishing of the treated water.

Sedimentation tanks called "secondary clarifiers" remove flocs of biological growth created in some methods of secondary treatment including activated sludge, trickling filters and rotating biological contactors.

Distillation

Distillation is a process of separating the component substances from a liquid mixture by selective evaporation and condensation. Distillation may result in essentially complete separation (nearly pure components), or it may be a partial separation that increases the concentration of selected components of the mixture. In either case the process exploits differences in the volatility of mixture's components. In industrial chemistry, distillation is a unit operation of practically universal importance, but it is a physical separation process and not a chemical reaction.

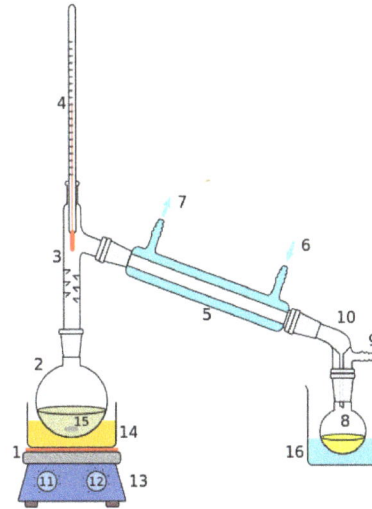

Laboratory display of distillation: **1**: A source of heat **2**: Still pot **3**: Still head **4**: Thermometer/Boiling point temperature **5**: Condenser **6**: Cooling water in **7**: Cooling water out **8**: Distillate/receiving flask **9**: Vacuum/gas inlet **10**: Still receiver **11**: Heat control **12**: Stirrer speed control **13**: Stirrer/heat plate **14**: Heating (Oil/sand) bath **15**: Stirring means e.g. (shown), boiling chips or mechanical stirrer **16**: Cooling bath.

Commercially, distillation has many applications. For example:

- In the fossil fuel industry distillation is a major class of operation in obtaining materials from crude oil for fuels and for chemical feedstocks.

- Distillation permits separation of air into its components — notably oxygen, nitrogen, and argon — for industrial use.

- In the field of industrial chemistry, large ranges of crude liquid products of chemical synthesis are distilled to separate them, either from other products, or from impurities, or from unreacted starting materials.

- Distillation of fermented products produces distilled beverages with a high alcohol content, or separates out other fermentation products of commercial value.

An installation for distillation, especially of alcohol, is a distillery. The distillation equipment is a still.

History

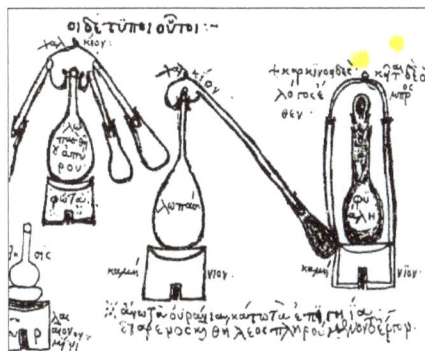

Distillation equipment used by the 3rd century Greek alchemist Zosimos of Panopolis, from the Byzantine Greek manuscript *Parisinus graces*.

Aristotle wrote about the process in his *Meteorologica* and even that "ordinary wine possesses a kind of exhalation, and that is why it gives out a flame". Later evidence of distillation comes from Greek alchemists working in Alexandria in the 1st century AD. Distilled water has been known since at least c. 200, when Alexander of Aphrodisias described the process. Distillation in China could have begun during the Eastern Han Dynasty (1st–2nd centuries), but archaeological evidence indicates that actual distillation of beverages began in the Jin and Southern Song dynasties. A still was found in an archaeological site in Qinglong, Hebei province dating to the 12th century. Distilled beverages were more common during the Yuan dynasty. Arabs learned the process from the Alexandrians and used it extensively in their chemical experiments.

Clear evidence of the distillation of alcohol comes from the School of Salerno in the 12th century. Fractional distillation was developed by Tadeo Alderotti in the 13th century.

In 1500, German alchemist Hieronymus Braunschweig published *Liber de arte destillandi* (The Book of the Art of Distillation) the first book solely dedicated to the subject of distillation, followed in 1512 by a much expanded version. In 1651, John French published The Art of Distillation the first major English compendium of practice, though it has been claimed that much of it derives from Braunschweig's work. This includes diagrams with people in them showing the industrial rather than bench scale of the operation.

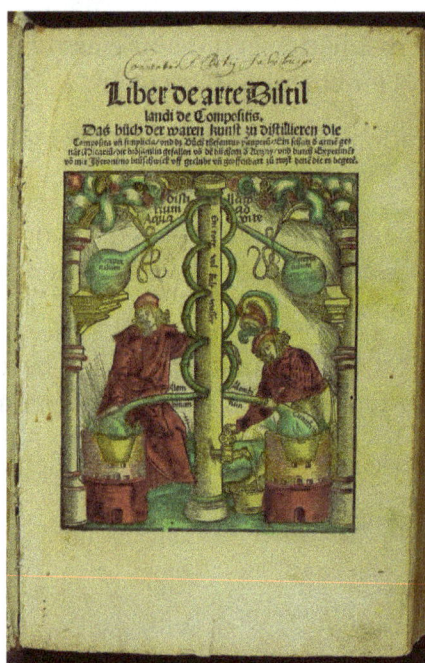

Hieronymus Brunschwig's *Liber de arte Distillandi de Compositis* (Strassburg, 1512) Chemical Heritage Foundation

A retort

Distillation

Old Ukrainian vodka still

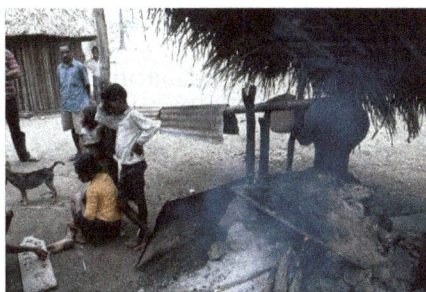

Simple liqueur distillation in East Timor

As alchemy evolved into the science of chemistry, vessels called retorts became used for distillations. Both alembics and retorts are forms of glassware with long necks pointing to the side at a downward angle which acted as air-cooled condensers to condense the distillate and let it drip downward for collection. Later, copper alembics were invented. Riveted joints were often kept tight by using various mixtures, for instance a dough made of rye flour. These alembics often featured a cooling system around the beak, using cold water for instance, which made the condensation of alcohol more efficient. These were called pot stills. Today, the retorts and pot stills have been largely supplanted by more efficient distillation methods in most industrial processes. However, the pot still is still widely used for the elaboration of some fine alcohols such as cognac, Scotch whisky, tequila and some vodkas. Pot stills made of various materials (wood, clay, stainless steel) are also used by bootleggers in various countries. Small pot stills are also sold for the domestic production of flower water or essential oils.

Early forms of distillation were batch processes using one vaporization and one condensation. Purity was improved by further distillation of the condensate. Greater volumes were processed by simply repeating the distillation. Chemists were reported to carry out as many as 500 to 600 distillations in order to obtain a pure compound.

In the early 19th century the basics of modern techniques including pre-heating and reflux were developed, particularly by the French, then in 1830 a British Patent was issued to Aeneas Coffey for a whisky distillation column, which worked continuously and may be regarded as the archetype of modern petrochemical units. In 1877, Ernest Solvay was granted a U.S. Patent for a tray column for ammonia distillation and the same and subsequent years saw developments of this theme for oil and spirits.

With the emergence of chemical engineering as a discipline at the end of the 19th century, scientific rather than empirical methods could be applied. The developing petroleum industry in the early 20th century provided the impetus for the development of accurate design methods such as the McCabe–Thiele method and the Fenske equation. The availability of powerful computers has also allowed direct computer simulation of distillation columns.

Applications of Distillation

The application of distillation can roughly be divided in four groups: laboratory scale, industrial distillation, distillation of herbs for perfumery and medicinals (herbal distillate), and food processing. The latter two are distinctively different from the former two in that in the processing of beverages and herbs, the distillation is not used as a true purification method but more to transfer all volatiles from the source materials to the distillate.

The main difference between laboratory scale distillation and industrial distillation is that laboratory scale distillation is often performed batch-wise, whereas industrial distillation often occurs continuously. In batch distillation, the composition of the source material, the vapors of the distilling compounds and the distillate change during the distillation. In batch distillation, a still is charged (supplied) with a batch of feed mixture, which is then separated into its component fractions which are collected sequentially from most volatile to less volatile, with the bottoms (remaining least or non-volatile fraction) removed at the end. The still can then be recharged and the process repeated.

In continuous distillation, the source materials, vapors, and distillate are kept at a constant composition by carefully replenishing the source material and removing fractions from both vapor and liquid in the system. This results in a better control of the separation process.

Idealized Distillation Model

The boiling point of a liquid is the temperature at which the vapor pressure of the liquid equals the pressure around the liquid, enabling bubbles to form without being crushed. A special case is the normal boiling point, where the vapor pressure of the liquid equals the ambient atmospheric pressure.

It is a common misconception that in a liquid mixture at a given pressure, each component boils at the boiling point corresponding to the given pressure and the vapors of each component will collect separately and purely. This, however, does not occur even in an idealized system. Idealized models of distillation are essentially governed by Raoult's law and Dalton's law, and assume that vapor–liquid equilibria are attained.

Raoult's law states that the vapor pressure of a solution is dependent on 1) the vapor pressure of each chemical component in the solution and 2) the fraction of solution each component makes up a.k.a. the mole fraction. This law applies to ideal solutions, or solutions that have different components but whose molecular interactions are the same as or very similar to pure solutions.

Dalton's law states that the total pressure is the sum of the partial pressures of each individual component in the mixture. When a multi-component liquid is heated, the vapor pressure of each component will rise, thus causing the total vapor pressure to rise. When the total vapor pressure reaches the pressure surrounding the liquid, boiling occurs and liquid turns to gas throughout the bulk of the liquid. Note that a mixture with a given composition has one boiling point at a given pressure, when the components are mutually soluble. A mixture of constant composition does not have multiple boiling points.

An implication of one boiling point is that lighter components never cleanly "boil first". At boiling point, all volatile components boil, but for a component, its percentage in the vapor is the same as its percentage of the total vapor pressure. Lighter components have a higher partial pressure and thus are concentrated in the vapor, but heavier volatile components also have a (smaller) partial pressure and necessarily evaporate also, albeit being less concentrated in the vapor. Indeed, batch distillation and fractionation succeed by varying the composition of the mixture. In batch distillation, the batch evaporates, which changes its composition; in fractionation, liquid higher in the fractionation column contains more lights and boils at lower temperatures. Therefore, starting from a given mixture, it appears to have a boiling range instead of a boiling *point*, although this is because its composition changes: each intermediate mixture has its own, singular boiling point.

The idealized model is accurate in the case of chemically similar liquids, such as benzene and toluene. In other cases, severe deviations from Raoult's law and Dalton's law are observed, most famously in the mixture of ethanol and water. These compounds, when heated together, form an azeotrope, which is a composition with a boiling point higher or lower than the boiling point of each separate liquid. Virtually all liquids, when mixed and heated, will display azeotropic behaviour. Although there are computational methods that can be used to estimate the behavior of a mixture of arbitrary components, the only way to obtain accurate vapor–liquid equilibrium data is by measurement.

It is not possible to *completely* purify a mixture of components by distillation, as this would require each component in the mixture to have a zero partial pressure. If ultra-pure products are the goal, then further chemical separation must be applied. When a binary mixture is evaporated and the other component, e.g. a salt, has zero partial pressure for practical purposes, the process is simpler and is called evaporation in engineering.

Batch Distillation

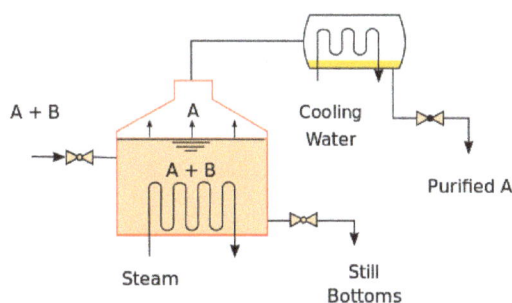

A batch still showing the separation of A and B.

Heating an ideal mixture of two volatile substances A and B (with A having the higher volatility, or lower boiling point) in a batch distillation setup (such as in an apparatus depicted in the opening figure) until the mixture is boiling results in a vapor above the liquid which contains a mixture of

A and B. The ratio between A and B in the vapor will be different from the ratio in the liquid: the ratio in the liquid will be determined by how the original mixture was prepared, while the ratio in the vapor will be enriched in the more volatile compound, A. The vapor goes through the condenser and is removed from the system. This in turn means that the ra-tio of compounds in the remaining liquid is now different from the initial ratio (i.e., more enriched in B than the starting liquid).

The result is that the ratio in the liquid mixture is changing, becoming richer in component B. This causes the boiling point of the mixture to rise, which in turn results in a rise in the temperature in the vapor, which results in a changing ratio of A : B in the gas phase (as distillation continues, there is an increasing proportion of B in the gas phase). This results in a slowly changing ratio A : B in the distillate.

If the difference in vapor pressure between the two components A and B is large (generally expressed as the difference in boiling points), the mixture in the beginning of the distillation is highly enriched in component A, and when component A has distilled off, the boiling liquid is enriched in component B.

Continuous Distillation

Continuous distillation is an ongoing distillation in which a liquid mixture is continuously (without interruption) fed into the process and separated fractions are removed continuously as output streams occur over time during the operation. Continuous distillation produces a minimum of two output fractions, including at least one volatile distillate fraction, which has boiled and been separately captured as a vapor, and then condensed to a liquid. There is always a bottoms (or residue) fraction, which is the least volatile residue that has not been separately captured as a condensed vapor.

Continuous distillation differs from batch distillation in the respect that concentrations should not change over time. Continuous distillation can be run at a steady state for an arbitrary amount of time. For any source material of specific composition, the main variables that affect the purity of products in continuous distillation are the reflux ratio and the number of theoretical equilibrium stages, in practice determined by the number of trays or the height of packing. Reflux is a flow from the condenser back to the column, which generates a recycle that allows a better separation with a given number of trays. Equilibrium stages are ideal steps where compositions achieve vapor–liquid equilibrium, repeating the separation process and allowing better separation given a reflux ratio. A column with a high reflux ratio may have fewer stages, but it refluxes a large amount of liquid, giving a wide column with a large holdup. Conversely, a column with a low reflux ratio must have a large number of stages, thus requiring a taller column.

General Improvements

Both batch and continuous distillations can be improved by making use of a fractionating column on top of the distillation flask. The column improves separation by providing a larger surface area for the vapor and condensate to come into contact. This helps it remain at equilibrium for as long as possible. The column can even consist of small subsystems ('trays' or 'dishes') which all contain an enriched, boiling liquid mixture, all with their own vapor–liquid equilibrium.

There are differences between laboratory-scale and industrial-scale fractionating columns, but the principles are the same. Examples of laboratory-scale fractionating columns (in increasing efficiency) include

- Air condenser

- Vigreux column (usually laboratory scale only)

- Packed column (packed with glass beads, metal pieces, or other chemically inert material)

- Spinning band distillation system.

Laboratory Scale Distillation

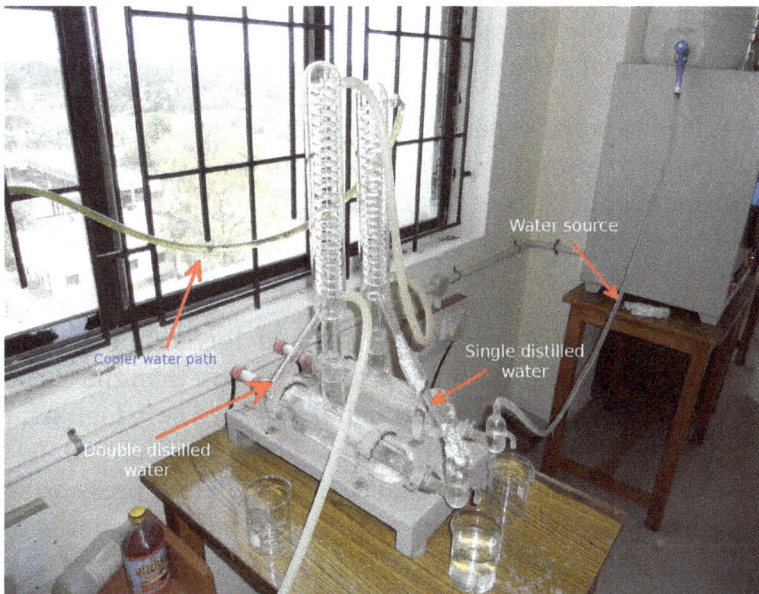

Typical laboratory distillation unit

Laboratory scale distillations are almost exclusively run as batch distillations. The device used in distillation, sometimes referred to as a *still*, consists at a minimum of a reboiler or pot in which the source material is heated, a condenser in which the heated vapour is cooled back to the liquid state, and a receiver in which the concentrated or purified liquid, called the distillate, is collected. Several laboratory scale techniques for distillation exist.

Simple Distillation

In simple distillation, the vapor is immediately channeled into a condenser. Consequently, the distillate is not pure but rather its composition is identical to the composition of the vapors at the given temperature and pressure. That concentration follows Raoult's law.

As a result, simple distillation is effective only when the liquid boiling points differ greatly (rule of thumb is 25 °C) or when separating liquids from non-volatile solids or oils. For these cases, the vapor pressures of the components are usually different enough that the distillate may be sufficiently pure for its intended purpose.

Fractional Distillation

For many cases, the boiling points of the components in the mixture will be sufficiently close that Raoult's law must be taken into consideration. Therefore, fractional distillation must be used in order to separate the components by repeated vaporization-condensation cycles within a packed fractionating column. This separation, by successive distillations, is also referred to as rectification.

As the solution to be purified is heated, its vapors rise to the fractionating column. As it rises, it cools, condensing on the condenser walls and the surfaces of the packing material. Here, the condensate continues to be heated by the rising hot vapors; it vaporizes once more. However, the composition of the fresh vapors are determined once again by Raoult's law. Each vaporization-condensation cycle (called a *theoretical plate*) will yield a purer solution of the more volatile component. In reality, each cycle at a given temperature does not occur at exactly the same position in the fractionating column; *theoretical plate* is thus a concept rather than an accurate description.

More theoretical plates lead to better separations. A spinning band distillation system uses a spinning band of Teflon or metal to force the rising vapors into close contact with the descending condensate, increasing the number of theoretical plates.

Steam Distillation

Like vacuum distillation, steam distillation is a method for distilling compounds which are heat-sensitive. The temperature of the steam is easier to control than the surface of a heating element, and allows a high rate of heat transfer without heating at a very high temperature. This process involves bubbling steam through a heated mixture of the raw material. By Raoult's law, some of the target compound will vaporize (in accordance with its partial pressure). The vapor mixture is cooled and condensed, usually yielding a layer of oil and a layer of water.

Dimethyl sulfoxide usually boils at 189 °C. Under a vacuum, it distills off into the receiver at only 70 °C.

Steam distillation of various aromatic herbs and flowers can result in two products; an essential oil as well as a watery herbal distillate. The essential oils are often used in perfumery and aromatherapy while the watery distillates have many applications in aromatherapy, food processing and skin care.

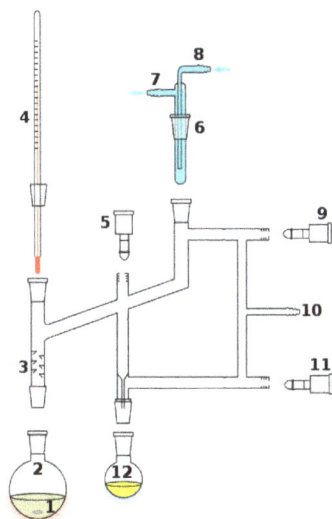

Perkin triangle distillation setup 1: Stirrer bar/anti-bumping granules **2:** Still pot **3:** Fractionating column **4:** Thermometer/Boiling point temperature **5:** Teflon tap 1 **6:** Cold finger **7:** Cooling water out **8:** Cooling water in **9:** Teflon tap 2 **10:** Vacuum/gas inlet **11:** Teflon tap 3 **12:** Still receiver

Vacuum Distillation

Some compounds have very high boiling points. To boil such compounds, it is often better to lower the pressure at which such compounds are boiled instead of increasing the temperature. Once the pressure is lowered to the vapor pressure of the compound (at the given temperature), boiling and the rest of the distillation process can commence. This technique is referred to as vacuum distillation and it is commonly found in the laboratory in the form of the rotary evaporator.

This technique is also very useful for compounds which boil beyond their decomposition temperature at atmospheric pressure and which would therefore be decomposed by any attempt to boil them under atmospheric pressure.

Molecular distillation is vacuum distillation below the pressure of 0.01 torr. 0.01 torr is one order of magnitude above high vacuum, where fluids are in the free molecular flow regime, i.e. the mean free path of molecules is comparable to the size of the equipment. The gaseous phase no longer exerts significant pressure on the substance to be evaporated, and consequently, rate of evaporation no longer depends on pressure. That is, because the continuum assumptions of fluid dynamics no longer apply, mass transport is governed by molecular dynamics rather than fluid dynamics. Thus, a short path between the hot surface and the cold surface is necessary, typically by suspending a hot plate covered with a film of feed next to a cold plate with a line of sight in between. Molecular distillation is used industrially for purification of oils.

Air-sensitive Vacuum Distillation

Some compounds have high boiling points as well as being air sensitive. A simple vacuum distillation system as exemplified above can be used, whereby the vacuum is replaced with an inert gas after the distillation is complete. However, this is a less satisfactory system if one desires to collect fractions under a reduced pressure. To do this a "cow" or "pig" adaptor can be added to the end of the condenser, or for better results or for very air sensitive compounds a Perkin triangle apparatus can be used.

The Perkin triangle, has means via a series of glass or Teflon taps to allows fractions to be isolated from the rest of the still, without the main body of the distillation being removed from either the vacuum or heat source, and thus can remain in a state of reflux. To do this, the sample is first isolated from the vacuum by means of the taps, the vacuum over the sample is then replaced with an inert gas (such as nitrogen or argon) and can then be stoppered and removed. A fresh collection vessel can then be added to the system, evacuated and linked back into the distillation system via the taps to collect a second fraction, and so on, until all fractions have been collected.

Short Path Distillation

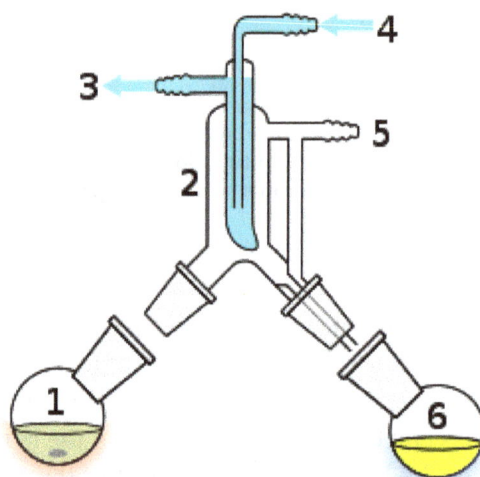

Short path vacuum distillation apparatus with vertical condenser (cold finger), to minimize the distillation path; **1:** Still pot with stirrer bar/anti-bumping granules **2:** Cold finger – bent to direct condensate **3:** Cooling water out **4:** cooling water in **5:** Vacuum/gas inlet **6:** Distillate flask/distillate.

Short path distillation is a distillation technique that involves the distillate travelling a short distance, often only a few centimeters, and is normally done at reduced pressure. A classic example would be a distillation involving the distillate travelling from one glass bulb to another, without the need for a condenser separating the two chambers. This technique is often used for compounds which are unstable at high temperatures or to purify small amounts of compound. The advantage is that the heating temperature can be considerably lower (at reduced pressure) than the boiling point of the liquid at standard pressure, and the distillate only has to travel a short distance before condensing. A short path ensures that little compound is lost on the sides of the apparatus. The Kugelrohr is a kind of a short path distillation apparatus which often contain multiple chambers to collect distillate fractions.

Zone Distillation

Zone distillation is a distillation process in long container with partial melting of refined matter in moving liquid zone and condensation of vapor in the solid phase at condensate pulling in cold area. The process is worked in theory. When zone heater is moving from the top to the bottom of the container then solid condensate with irregular impurity distribution is forming. Then most pure part of the condensate may be extracted as product. The process may be iterated many times by moving (without turnover) the received condensate to the bottom part of the container on the place of refined matter. The irregular impurity distribution in the condensate (that is efficiency of purification) increases with number of repetitions of the process. Zone distillation is a distillation analog of zone recrystallization. Impurity distribution in the condensate is described by known equations of zone recrystallization with various numbers of iteration of process – with replacement distribution efficient k of crystallization on separation factor α of distillation.

Other Types

- The process of reactive distillation involves using the reaction vessel as the still. In this process, the product is usually significantly lower-boiling than its reactants. As the product is formed from the reactants, it is vaporized and removed from the reaction mixture. This technique is an example of a continuous vs. a batch process; advantages include less downtime to charge the reaction vessel with starting material, and less workup. Distillation "over a reactant" could be classified as a reactive distillation. It is typically used to remove volatile impurity from the distallation feed. For example, a little lime may be added to remove carbon dioxide from water followed by a second distillation with a little sulphuric acid added to remove traces of ammonia.

- Catalytic distillation is the process by which the reactants are catalyzed while being distilled to continuously separate the products from the reactants. This method is used to assist equilibrium reactions reach completion.

- Pervaporation is a method for the separation of mixtures of liquids by partial vaporization through a non-porous membrane.

- Extractive distillation is defined as distillation in the presence of a miscible, high boiling, relatively non-volatile component, the solvent, that forms no azeotrope with the other components in the mixture.

- Flash evaporation (or partial evaporation) is the partial vaporization that occurs when a saturated liquid stream undergoes a reduction in pressure by passing through a throttling valve or other throttling device. This process is one of the simplest unit operations, being equivalent to a distillation with only one equilibrium stage.

- Codistillation is distillation which is performed on mixtures in which the two compounds are not miscible.

The unit process of evaporation may also be called "distillation":

- In rotary evaporation a vacuum distillation apparatus is used to remove bulk solvents from a sample. Typically the vacuum is generated by a water aspirator or a membrane pump.

- In a kugelrohr a short path distillation apparatus is typically used (generally in combination with a (high) vacuum) to distill high boiling (> 300 °C) compounds. The apparatus consists of an oven in which the compound to be distilled is placed, a receiving portion which is outside of the oven, and a means of rotating the sample. The vacuum is normally generated by using a high vacuum pump.

Other uses:

- Dry distillation or destructive distillation, despite the name, is not truly distillation, but rather a chemical reaction known as pyrolysis in which solid substances are heated in an inert or reducing atmosphere and any volatile fractions, containing high-boiling liquids and products of pyrolysis, are collected. The destructive distillation of wood to give methanol is the root of its common name – *wood alcohol*.

- Freeze distillation is an analogous method of purification using freezing instead of evaporation. It is not truly distillation, but a recrystallization where the product is the mother liquor, and does not produce products equivalent to distillation. This process is used in the production of ice beer and ice wine to increase ethanol and sugar content, respectively. It is also used to produce applejack. Unlike distillation, freeze distillation concentrates poisonous congeners rather than removing them; As a result, many countries prohibit such applejack as a health measure. However, reducing methanol with the absorption of 4A molecular sieve is a practical method for production. Also, distillation by evaporation can separate these since they have different boiling points.

Azeotropic Distillation

Interactions between the components of the solution create properties unique to the solution, as most processes entail nonideal mixtures, where Raoult's law does not hold. Such interactions can result in a constant-boiling azeotrope which behaves as if it were a pure compound (i.e., boils at a single temperature instead of a range). At an azeotrope, the solution contains the given component in the same proportion as the vapor, so that evaporation does not change the purity, and distillation does not effect separation. For example, ethyl alcohol and water form an azeotrope of 95.6% at 78.1 °C.

If the azeotrope is not considered sufficiently pure for use, there exist some techniques to break the azeotrope to give a pure distillate. This set of techniques are known as azeotropic distillation. Some techniques achieve this by "jumping" over the azeotropic composition (by adding another component to create a new azeotrope, or by varying the pressure). Others work by chemically or physically removing or sequestering the impurity. For example, to purify ethanol beyond 95%, a drying agent (or desiccant, such as potassium carbonate) can be added to convert the soluble water into insoluble water of crystallization. Molecular sieves are often used for this purpose as well.

Immiscible liquids, such as water and toluene, easily form azeotropes. Commonly, these azeotropes are referred to as a low boiling azeotrope because the boiling point of the azeotrope is lower than the boiling point of either pure component. The temperature and composition of the azeotrope is easily predicted from the vapor pressure of the pure components, without use of Raoult's law. The azeotrope is easily broken in a distillation set-up by using a liquid–liquid separator (a decanter) to separate the two liquid layers that are condensed overhead. Only one of the two liquid layers is refluxed to the distillation set-up.

High boiling azeotropes, such as a 20 weight percent mixture of hydrochloric acid in water, also exist. As implied by the name, the boiling point of the azeotrope is greater than the boiling point of either pure component.

To break azeotropic distillations and cross distillation boundaries, such as in the DeRosier Problem, it is necessary to increase the composition of the light key in the distillate.

Breaking an Azeotrope with Unidirectional Pressure Manipulation

The boiling points of components in an azeotrope overlap to form a band. By exposing an azeotrope to a vacuum or positive pressure, it's possible to bias the boiling point of one component away from the other by exploiting the differing vapour pressure curves of each; the curves may overlap at the azeotropic point, but are unlikely to be remain identical further along the pressure axis either side of the azeotropic point. When the bias is great enough, the two boiling points no longer overlap and so the azeotropic band disappears.

This method can remove the need to add other chemicals to a distillation, but it has two potential drawbacks.

Under negative pressure, power for a vacuum source is needed and the reduced boiling points of the distillates requires that the condenser be run cooler to prevent distillate vapours being lost to the vacuum source. Increased cooling demands will often require additional energy and possibly new equipment or a change of coolant.

Alternatively, if positive pressures are required, standard glassware can not be used, energy must be used for pressurization and there is a higher chance of side reactions occurring in the distillation, such as decomposition, due to the higher temperatures required to effect boiling.

A unidirectional distillation will rely on a pressure change in one direction, either positive or negative.

Pressure-swing Distillation

Pressure-swing distillation is essentially the same as the unidirectional distillation used to break azeotropic mixtures, but here both positive and negative pressures may be employed.

This improves the selectivity of the distillation and allows a chemist to optimize distillation by avoiding extremes of pressure and temperature that waste energy. This is particularly important in commercial applications.

One example of the application of pressure-swing distillation is during the industrial purification of ethyl acetate after its catalytic synthesis from ethanol.

Industrial Distillation

Large scale industrial distillation applications include both batch and continuous fractional, vacuum, azeotropic, extractive, and steam distillation. The most widely used industrial applications of continuous, steady-state fractional distillation are in petroleum refineries, petrochemical and chemical plants and natural gas processing plants.

Typical industrial distillation towers

To control and optimize such industrial distillation, a standardized laboratory method, ASTM D86, is established. This test method extends to the atmospheric distillation of petroleum products using a laboratory batch distillation unit to quantitatively determine the boiling range characteristics of petroleum products.

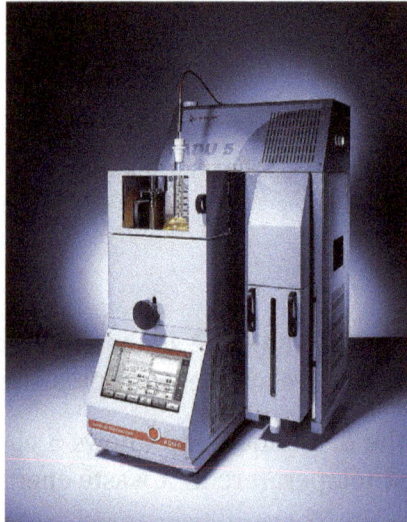

Automatic Distillation Unit for the determination of the boiling range of petroleum products at atmospheric pressure

Industrial distillation is typically performed in large, vertical cylindrical columns known as distillation towers or distillation columns with diameters ranging from about 65 centimeters to 16 meters and heights ranging from about 6 meters to 90 meters or more. When the process feed has a diverse composition, as in distilling crude oil, liquid outlets at intervals up the column allow for the withdrawal of different *fractions* or products having different boiling points or boiling ranges. The "lightest" products (those with the lowest boiling point) exit from the top of the columns and the "heaviest" products (those with the highest boiling point) exit from the bottom of the column and are often called the bottoms.

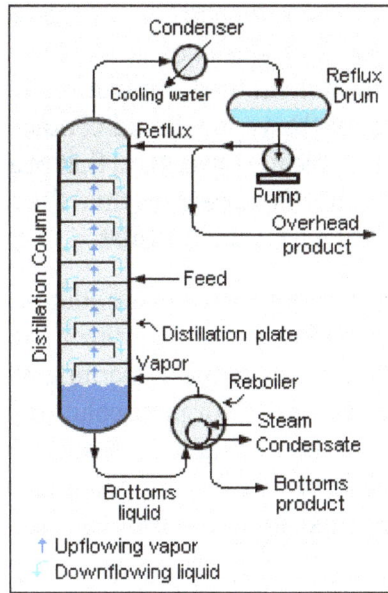

Diagram of a typical industrial distillation tower

Industrial towers use reflux to achieve a more complete separation of products. Reflux refers to the portion of the condensed overhead liquid product from a distillation or fractionation tower that is returned to the upper part of the tower as shown in the schematic diagram of a typical, large-scale industrial distillation tower. Inside the tower, the downflowing reflux liquid provides cooling and condensation of the upflowing vapors thereby increasing the efficiency of the distillation tower. The more reflux that is provided for a given number of theoretical plates, the better the tower's separation of lower boiling materials from higher boiling materials. Alternatively, the more reflux that is provided for a given desired separation, the fewer the number of theoretical plates required. Chemical engineers must choose what combination of reflux rate and number of plates is both economically and physically feasible for the products purified in the distillation column.

Such industrial fractionating towers are also used in cryogenic air separation, producing liquid oxygen, liquid nitrogen, and high purity argon. Distillation of chlorosilanes also enables the production of high-purity silicon for use as a semiconductor.

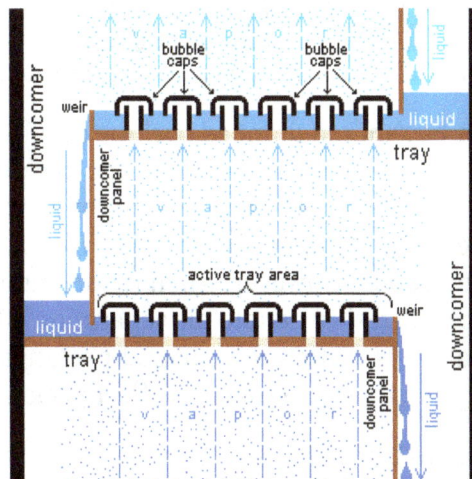

Section of an industrial distillation tower showing detail of trays with bubble caps

Design and operation of a distillation tower depends on the feed and desired products. Given a simple, binary component feed, analytical methods such as the McCabe–Thiele method or the Fenske equation can be used. For a multi-component feed, simulation models are used both for design and operation. Moreover, the efficiencies of the vapor–liquid contact devices (referred to as "plates" or "trays") used in distillation towers are typically lower than that of a theoretical 100% efficient equilibrium stage. Hence, a distillation tower needs more trays than the number of theoretical vapor–liquid equilibrium stages. A variety of models have been postulated to estimate tray efficiencies.

In modern industrial uses, a packing material is used in the column instead of trays when low pressure drops across the column are required. Other factors that favor packing are: vacuum systems, smaller diameter columns, corrosive systems, systems prone to foaming, systems requiring low liquid holdup, and batch distillation. Conversely, factors that favor plate columns are: presence of solids in feed, high liquid rates, large column diameters, complex columns, columns with wide feed composition variation, columns with a chemical reaction, absorption columns, columns limited by foundation weight tolerance, low liquid rate, large turn-down ratio and those processes subject to process surges.

Large-scale, industrial vacuum distillation column

This packing material can either be random dumped packing (1–3" wide) such as Raschig rings or structured sheet metal. Liquids tend to wet the surface of the packing and the vapors pass across this wetted surface, where mass transfer takes place. Unlike conventional tray distillation in which every tray represents a separate point of vapor–liquid equilibrium, the vapor–liquid equilibrium curve in a packed column is continuous. However, when modeling packed columns, it is useful to compute a number of "theoretical stages" to denote the separation efficiency of the packed column with respect to more traditional trays. Differently shaped packings have different surface areas and void space between packings. Both of these factors affect packing performance.

Another factor in addition to the packing shape and surface area that affects the performance of random or structured packing is the liquid and vapor distribution entering the packed bed. The number of theoretical stages required to make a given separation is calculated using a specific vapor to liquid ratio. If the liquid and vapor are not evenly distributed across the superficial tower area as it enters the packed bed, the liquid to vapor ratio will not be correct in the packed bed and the required separation will not be achieved. The packing will appear to not be working properly. The height equivalent to a theoretical plate (HETP) will be greater than expected. The problem is not the packing itself but the mal-distribution of the fluids entering the packed bed. Liquid mal-distribution is more frequently the problem than vapor. The design of the liquid distributors used to introduce the feed and reflux to a packed bed is critical to making the packing perform to it maximum efficiency. Methods of evaluating the effectiveness of a liquid distributor to evenly distribute the liquid entering a packed bed can be found in references. Considerable work as been done on this topic by Fractionation Research, Inc. (commonly known as FRI).

Multi-effect Distillation

The goal of multi-effect distillation is to increase the energy efficiency of the process, for use in desalination, or in some cases one stage in the production of ultrapure water. The number of effects is inversely proportional to the kW·h/m³ of water recovered figure, and refers to the volume of water recovered per unit of energy compared with single-effect distillation. One effect is roughly 636 kW·h/m³.

- Multi-stage flash distillation Can achieve more than 20 effects with thermal energy input, as mentioned in the article.

- Vapor compression evaporation Commercial large-scale units can achieve around 72 effects with electrical energy input, according to manufacturers.

There are many other types of multi-effect distillation processes, including one referred to as simply multi-effect distillation (MED), in which multiple chambers, with intervening heat exchangers, are employed.

Distillation in food Processing

Distilled Beverages

Carbohydrate-containing plant materials are allowed to ferment, producing a dilute solution of ethanol in the process. Spirits such as whiskey and rum are prepared by distilling these dilute solutions of ethanol. Components other than ethanol, including water, esters, and other alcohols, are collected in the condensate, which account for the flavor of the beverage. Some of these beverages are then stored in barrels or other containers to acquire more flavor compounds and characteristic flavors.

Water Chlorination

Water chlorination is the process of adding chlorine (Cl_2) or hypochlorite to water. This method is used to kill certain bacteria and other microbes in tap water as chlorine is highly toxic. In particular, chlorination is used to prevent the spread of waterborne diseases such as cholera, dysentery, typhoid etc.

History

In a paper published in 1894, it was formally proposed to add chlorine to water to render it "germ-free". Two other authorities endorsed this proposal and published it in many other papers in 1895. Early attempts at implementing water chlorination at a water treatment plant were made in 1893 in Hamburg, Germany, and in 1897 the town of Maidstone, England was the first to have its entire water supply treated with chlorine.

Permanent water chlorination began in 1905, when a faulty slow sand filter and a contaminated water supply caused a serious typhoid fever epidemic in Lincoln, England. Dr. Alexander Cruickshank Houston used chlorination of the water to stop the epidemic. His installation fed a concentrated solution of so-called *chloride of lime* to the water being treated. This was not simply modern calcium chloride, but contained chlorine gas dissolved in lime-water (dilute calcium hydroxide) to form calcium hypochlorite (chlorinated lime). The chlorination of the water supply helped stop the epidemic and as a precaution, the chlorination was continued until 1911 when a new water supply was instituted.

The first continuous use of chlorine in the United States for disinfection took place in 1908 at Boonton Reservoir (on the Rockaway River), which served as the supply for Jersey City, New Jersey. Chlorination was achieved by controlled additions of dilute solutions of chloride of lime (calcium hypochlorite) at doses of 0.2 to 0.35 ppm. The treatment process was conceived by Dr. John L. Leal, and the chlorination plant was designed by George Warren Fuller. Over the next few years, chlorine disinfection using chloride of lime (calcium hypochlorite) were rapidly installed in drinking water systems around the world.

The technique of purification of drinking water by use of compressed liquefied chlorine gas was developed by a British officer in the Indian Medical Service, Vincent B. Nesfield, in 1903. According to his own account, "It occurred to me that chlorine gas might be found satisfactory ... if suitable means could be found for using it.... The next important question was how to render the gas portable. This might be accomplished in two ways: By liquefying it, and storing it in lead-lined iron vessels, having a jet with a very fine capillary canal, and fitted with a tap or a screw cap. The tap is turned on, and the cylinder placed in the amount of water required. The chlorine bubbles out, and in ten to fifteen minutes the water is absolutely safe. This method would be of use on a large scale, as for service water carts."

Major Carl Rogers Darnall, Professor of Chemistry at the Army Medical School, gave the first practical demonstration of this in 1910. This work became the basis for present day systems of municipal water *purification*. Shortly after Darnall's demonstration, Major William J. L. Lyster of the Army Medical Department used a solution of calcium hypochlorite in a linen bag to treat water. For many decades, Lyster's method remained the standard for U.S. ground forces in the field and in camps, implemented in the form of the familiar Lyster Bag (also spelled Lister Bag).

Chlorine gas was first used on a continuing basis to disinfect the water supply at the Belmont filter plant, Philadelphia, Pennsylvania by using a machine invented by Charles Frederick Wallace who dubbed it the Chlorinator. It was manufactured by the Wallace & Tiernan company beginning in 1913. By 1941, disinfection of U.S. drinking water by chlorine gas had largely replaced the use of chloride of lime.

Chlorination can also be practiced using sodium hypochlorite or various other chemicals.

Biochemistry

As a halogen, chlorine is a highly efficient disinfectant, and is added to public water supplies to kill disease-causing pathogens, such as bacteria, viruses, and protozoans, that commonly grow in water supply reservoirs, on the walls of water mains and in storage tanks. The microscopic agents of many diseases such as cholera, typhoid fever, and dysentery killed countless people annually before disinfection methods were employed routinely.

Chlorine is manufactured from salt by electrolysis or other methods. It is a gas at atmospheric pressures but liquefies under pressure. The liquefied gas is transported and used as such.

As a strong oxidizing agent, chlorine kills via the oxidation of organic molecules. Chlorine and hydrolysis product hypochlorous acid are neutrally charged and therefore easily penetrate the negatively charged surface of pathogens. It is able to disintegrate the lipids that compose the cell wall and react with intracellular enzymes and proteins, making them nonfunctional. Microorganisms then either die or are no longer able to multiply.

Principles

When dissolved in water, chlorine converts to an equilibrium mixture of chlorine, hypochlorous acid (HOCl), and hydrochloric acid (HCl):

$$Cl_2 + H_2O \rightleftharpoons HOCl + HCl$$

In acidic solution, the major species are Cl2 and HOCl, whereas in alkaline solution, effectively only ClO^- (hypochlorite ion) is present. Very small concentrations of ClO_2^-, ClO_3^-, ClO_4^- are also found.

Shock Chlorination

Shock chlorination is a process used in many swimming pools, water wells, springs, and other water sources to reduce the bacterial and algal residue in the water. Shock chlorination is performed by mixing a large amount of hypochlorite into the water. The hypochlorite can be in the form of a powder or a liquid such as chlorine bleach (solution of sodium hypochlorite in water). Water that is being shock chlorinated should not be swum in or drunk until the sodium hypochlorite count in the water goes down to three parts per million (PPM) or less.

Drawbacks to Water Chlorination

Disinfection by chlorination can be problematic, in some circumstances. Chlorine can react with naturally occurring organic compounds found in the water supply to produce compounds known as disinfection by-products (DBPs). The most common DBPs are trihalomethanes (THMs) and haloacetic acids (HAAs). Trihalomethanes are the main disinfectant by-products created from chlorination with two different types, bromoform and dibromochloromethane, which are mainly responsible for health hazards. Their effects depend strictly on the duration of their exposure to the chemicals and the amount ingested into the body. In high doses, bromoform mainly slows down regular brain activity, which is manifested by symptoms such as sleepiness or sedation. Chronic exposure of both bromoform and dibromochloromethane can cause liver and kidney cancer, as well as heart disease, unconsciousness, or death in high doses. Due to the potential carcinogenicity of these compounds, drinking water regulations across the developed world require regular monitoring of the concentration of these compounds in the distribution systems of municipal water systems. The World Health Organization has stated that "the risks to health from these by-products are extremely small in comparison with the risks associated with inadequate disinfection".

There are also other concerns regarding chlorine, including its volatile nature which causes it to disappear too quickly from the water system, and organoleptic concerns such as taste and odor.

Alternative Methods for Water Disinfection

Ozonation

Ozonation is used by many European countries and also in a few municipalities in the United States and Canada. This alternative is more cost effective and energy intensive. It involves ozone being bubbled through the water, breaking down all parasites, bacteria, and all other harmful organic substances. However, this method leaves no residual ozone to control contamination of the water after the process has been completed.

The advantage of chlorine in comparison to ozone is that the residual persists in the water for an extended period of time. This feature allows the chlorine to travel through the water supply system, effectively controlling pathogenic backflow contamination. In a large system this may not be adequate, and so chlorine levels may be boosted at points in the distribution system, or chloramine may be used, which remains in the water for longer before reacting or dissipating.

Chloramination

Chloramination is also becoming increasingly common. Disinfection with chloramine produces less undesirable byproducts than chlorine (gas or hypochlorite). Chloramine has a longer half-life in the distribution system, and maintains effective protection against pathogens. Chloramines persist in the distribution because of their lower redox potential in comparison to free chlorine. Chloramine is formed by adding ammonia and chlorine into drinking water to form monochloramine and/or dichloramine. Whereas *Helicobacter pylori* can be many times more resistant to chlorine than *Escherichia coli*, both organisms are about equally susceptible to the disinfecting effect of chloramine.

Bromination and Iodinization

Chlorine in water is over three times more effective as a disinfectant against *Escherichia coli* than an equivalent concentration of bromine, and over six times more effective than an equivalent concentration of iodine.

Home Filtration

Water treated by filtration and home filtration may not need further disinfection; a very high proportion of pathogens are removed by materials in the filter bed. Filtered water must be used soon after it is filtered, as the low amount of remaining microbes may proliferate over time. In general, these home filters remove over 90% of the chlorine available to a glass of treated water. These filters must be periodically replaced otherwise the bacterial content of the water may actually increase due to the growth of bacteria within the filter unit.

UV Radiation

UV disinfection is gaining popularity. UV treatment leaves minimal residue in the water. In water UV generates ozone in situ and thus has many of the advantages of ozone disinfection. However, ultraviolet germicidal irradiation alone (as well as chlorination alone) will not remove toxins from bacteria, pesticides, heavy metals, etc. from water.

Ionizing Radiation

Like UV, ionizing radiation (X-rays, gamma rays, and electron beams) has been used to sterilize water.

References

- Chen, Jimmy, and Regli, Stig. (2002). "Disinfection Practices and Pathogen Inactivation in ICR Surface Water Plants." Information Collection Rule Data Analysis. Denver:American Water Works Association. McGuire, Michael J., McLain, Jennifer L. and Obolensky, Alexa, eds. pp. 376–378. ISBN 1-58321-273-6

- Crittenden, John C., et al., eds. (2005). Water Treatment: Principles and Design. 2nd Edition. Hoboken, NJ:Wiley. ISBN 0-471-11018-3

- Kawamura, Susumu. (2000). Integrated Design and Operation of Water Treatment Facilities. 2nd Edition. New York:Wiley. pp. 74–5, 104. ISBN 0-471-35093-1

- Gunn, S. William A. & Masellis, Michele (2007). Concepts and Practice of Humanitarian Medicine. Springer. p. 87. ISBN 978-0-387-72264-1.

- Franson, Mary Ann. Standard Methods for the Examination of Water and Wastewater. 14th ed. (1975) APHA, AWWA & WPCF. ISBN 0-87553-078-8. pp. 89–98

- Harwood, Laurence M.; Moody, Christopher J. (1989). Experimental organic chemistry: Principles and Practice (Illustrated ed.). Oxford: Blackwell Scientific Publications. pp. 141–143. ISBN 978-0-632-02017-1.

- Bryan H. Bunch; Alexander Hellemans (2004). The History of Science and Technology. Houghton Mifflin Harcourt. p. 88. ISBN 0-618-22123-9.

- Forbes, Robert James (1970). A short history of the art of distillation: from the beginnings up to the death of Cellier Blumenthal. BRILL. pp. 57, 89. ISBN 978-90-04-00617-1. Retrieved 29 June 2010.

- D. F. Othmer (1982) "Distillation – Some Steps in its Development", in W. F. Furter (ed) A Century of Chemical Engineering ISBN 0-306-40895-3

- Goldman, Steven J., Jackson, Katharine & Bursztynsky, Taras A. Erosion & Sediment Control Handbook. McGraw-Hill (1986). ISBN 0-07-023655-0.

7

Waterborne Diseases

Pathogenic microorganisms that are most commonly transmitted by contaminated fresh water cause waterborne diseases. The waterborne diseases that are dealt within this chapter are amoebiasism, cryptosporidiosis, schistosomiasis and enterobiasis. This chapter is a compilation of the various branches of water pollution and management that form an integral part of the broader subject matter.

Amoebiasis

Amoebiasis, also known as amebiasis or entamoebiasis, is an infection caused by any of the amoebas of the *Entamoeba* group. Symptoms are most common upon infection by *Entamoeba histolytica*. Amoebiasis can present with no, mild, or severe symptoms. Symptoms may include abdominal pain, mild diarrhoea, bloody diarrhea or severe colitis with tissue death and perforation. This last complication may cause peritonitis. People affected may develop anemia due to loss of blood.

Invasion of the intestinal lining causes amoebic bloody diarrhea or amoebic colitis. If the parasite reaches the bloodstream it can spread through the body, most frequently ending up in the liver where it causes amoebic liver abscesses. Liver abscesses can occur without previous diarrhea. Cysts of *Entamoeba* can survive for up to a month in soil or for up to 45 minutes under fingernails. It is important to differentiate between amoebiasis and bacterial colitis. The preferred diagnostic method is through faecal examination under microscope, but requires a skilled microscopist and may not be reliable when excluding infection. This method however may not be able to separate between specific types. Increased white blood cell count is present in severe cases, but not in mild ones. The most accurate test is for antibodies in the blood, but it may remain positive following treatment.

Prevention of amoebiasis is by separating food and water from faeces and by proper sanitation measures. There is no vaccine. There are two treatment options depending on the location of the infection. Amoebiasis in tissues is treated with either metronidazole, tinidazole, nitazoxanide, dehydroemetine or chloroquine, while luminal infection is treated with diloxanide furoate or iodoquinoline. For treatment to be effective against all stages of the amoeba may require a combination of medications. Infections without symptoms do not require treatment but infected individuals can spread the parasite to others and treatment can be considered. Treatment of other *Entamoeba* infections apart from *E. histolytica* is not needed.

Amoebiasis is present all over the world. About 480 million people are infected with what appears to be *E. histolytica* and these result in the death of between 40,000–110,000 people every year. Most infections are now ascribed to *E. dispar*. *E. dispar* is more common in certain areas and symptomatic cases may be fewer than previously reported. The first case of amoebiasis was documented in 1875 and in 1891 the disease was described in detail, resulting in the terms *amoebic dysentery* and *amoebic liver abscess*. Further evidence from the Philippines in 1913 found that upon ingesting cysts of *E. histolytica* volunteers developed the disease. It has been known since 1897 that at least one non-disease-causing species of *Entamoeba* existed (*Entamoeba coli*), but it was first formally recognized by the WHO in 1997 that *E. histolytica* was two species, despite this having first been proposed in 1925. In addition to the now-recognized *E. dispar* evidence shows there are at least two other species of *Entamoeba* that look the same in humans - *E. moshkovskii* and *Entamoeba bangladeshi*. The reason these species haven't been differentiated until recently is because of the reliance on appearance.

Signs and Symptoms

Most infected people, about 90%, are asymptomatic, but this disease has the potential to make the sufferer dangerously ill. It is estimated that about 40,000 to 100,000 people worldwide die annually due to amoebiasis.

Infections can sometimes last for years. Symptoms take from a few days to a few weeks to develop and manifest themselves, but usually it is about two to four weeks. Symptoms can range from mild diarrhea to severe dysentery with blood and mucus. The blood comes from lesions formed by the amoebae invading the lining of the large intestine. In about 10% of invasive cases the amoebae enter the bloodstream and may travel to other organs in the body. Most commonly this means the liver, as this is where blood from the intestine reaches first, but they can end up almost anywhere in the body.

Onset time is highly variable and the average asymptomatic infection persists for over a year. It is theorized that the absence of symptoms or their intensity may vary with such factors as strain of amoeba, immune response of the host, and perhaps associated bacteria and viruses.

In asymptomatic infections the amoeba lives by eating and digesting bacteria and food particles in the gut, a part of the gastrointestinal tract. It does not usually come in contact with the intestine itself due to the protective layer of mucus that lines the gut. Disease occurs when amoeba comes in contact with the cells lining the intestine. It then secretes the same substances it uses to digest bacteria, which include enzymes that destroy cell membranes and proteins. This process can lead to penetration and digestion of human tissues, resulting first in flask-shaped ulcers in the intestine. *Entamoeba histolytica* ingests the destroyed cells by phagocytosis and is often seen with red blood cells (a process known as erythrophagocytosis) inside when viewed in stool samples. Especially in Latin America, a granulomatous mass (known as an amoeboma) may form in the wall of the ascending colon or rectum due to long-lasting immunological cellular response, and is sometimes confused with cancer.

"Theoretically, the ingestion of one viable cyst can cause an infection."

Cause

Amoebiasis is an infection caused by the amoeba *Entamoeba histolytica*. Likewise amoebiasis is sometimes incorrectly used to refer to infection with other amoebae, but strictly speaking it should be reserved for *Entamoeba histolytica* infection. Other amoebae infecting humans include:

- Parasites

    o *Dientamoeba fragilis*, which causes Dientamoebiasis

o   *Entamoeba dispar*

o   *Entamoeba hartmanni*

o   *Entamoeba coli*

o   *Entamoeba moshkovskii*

o   *Endolimax nana* and

o   *Iodamoeba butschlii*.

Except for *Dientamoeba*, the parasites above are not thought to cause disease.

- Free living amoebas. These species are often described as "opportunistic free-living amoebas" as human infection is not an obligate part of their life cycle.

  o   *Naegleria fowleri*, which causes Primary amoebic meningoencephalitis

  o   *Acanthamoeba*, which causes Cutaneous amoebiasis and Acanthamoeba keratitis

  o   *Balamuthia mandrillaris*, which causes Granulomatous amoebic encephalitis and Primary amoebic meningoencephalitis

  o   *Sappinia diploidea*

Transmission

Amoebiasis is usually transmitted by the fecal-oral route, but it can also be transmitted indirectly through contact with dirty hands or objects as well as by anal-oral contact. Infection is spread through ingestion of the cyst form of the parasite, a semi-dormant and hardy structure found in feces. Any non-encysted amoebae, or *trophozoites*, die quickly after leaving the body but may also be present in stool: these are rarely the source of new infections. Since amoebiasis is transmitted through contaminated food and water, it is often endemic in regions of the world with limited modern sanitation systems, including México, Central America, western South America, South Asia, and western and southern Africa.

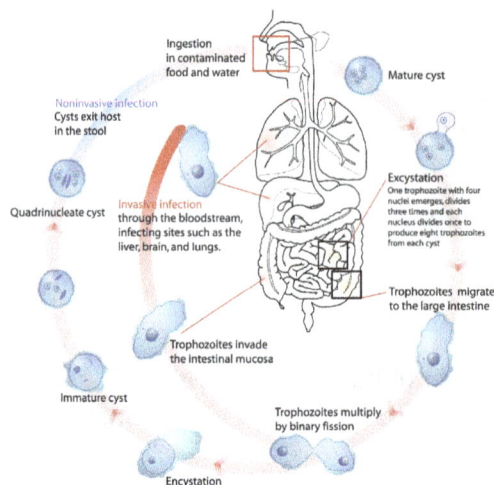

Life-cycle of the *Entamoeba histolytica*

Amoebic dysentery is often confused with "traveler's diarrhea" because of its prevalence in developing nations. In fact, most traveler's diarrhea is bacterial or viral in origin.

Diagnosis

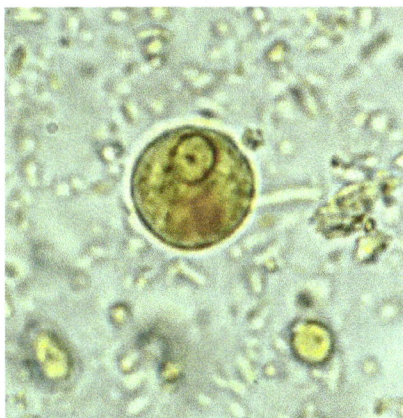

Immature *E. histolytica/E. dispar* cyst in a concentrated wet mount stained with iodine. This early cyst has only one nucleus and a glycogen mass is visible (brown stain). From CDC's Division of Parasitic Diseases With colonoscopy it is possible to detect small ulcers of between 3–5mm, but diagnosis may be difficult as the mucous membrane between these areas can look either healthy or inflamed.

Asymptomatic human infections are usually diagnosed by finding cysts shed in the stool. Various flotation or sedimentation procedures have been developed to recover the cysts from fecal matter and stains help to visualize the isolated cysts for microscopic examination. Since cysts are not shed constantly, a minimum of three stools should be examined. In symptomatic infections, the motile form (the trophozoite) can often be seen in fresh feces. Serological tests exist and most individuals (whether with symptoms or not) will test positive for the presence of antibodies. The levels of antibody are much higher in individuals with liver abscesses. Serology only becomes positive about two weeks after infection. More recent developments include a kit that detects the presence of amoeba proteins in the feces and another that detects ameba DNA in feces. These tests are not in widespread use due to their expense.

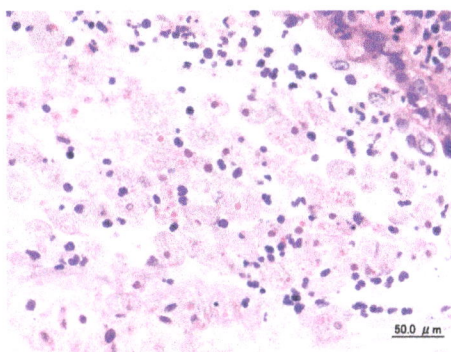

Amoebae in a colon biopsy from a case of amoebic dysentery.

Microscopy is still by far the most widespread method of diagnosis around the world. However it is not as sensitive or accurate in diagnosis as the other tests available. It is important to distinguish the *E. histolytica* cyst from the cysts of nonpathogenic intestinal protozoa such as *Entamoeba coli* by its appearance. *E. histolytica* cysts have a maximum of four nuclei, while the commensal *Entamoeba coli* cyst has up to 8 nuclei. Additionally, in *E. histolytica,* the endosome is centrally located in the nucleus, while it is usually off-center in *Entamoeba coli*. Finally, chromatoidal bodies in *E. histolytica* cysts are rounded, while they are jagged in *Entamoeba coli*. However, other species, *Entamoeba dispar* and *E. moshkovskii*, are also commensals and cannot be distinguished from *E. histolytica* under the microscope. As *E. dispar* is much more common than *E. histolytica* in most parts of the world this means that there is a lot of incorrect diagnosis of *E. histolytica* infection taking place. The WHO recommends that infections diagnosed by microscopy alone should not be treated if they are asymptomatic and there is no other reason to suspect that the infection is actually *E. histolytica*. Detection of cysts or trophozoites stools under microscope may require examination of several samples over several days to determine if they are present, because cysts are shed intermittently and may not show up in every sample.

Typically, the organism can no longer be found in the feces once the disease goes extra-intestinal. Serological tests are useful in detecting infection by *E. histolytica* if the organism goes extra-intestinal and in excluding the organism from the diagnosis of other disorders. An Ova & Parasite (O&P) test or an *E. histolytica* fecal antigen assay is the proper assay for intestinal infections. Since antibodies may persist for years after clinical cure, a positive serological result may not necessarily indicate an active infection. A negative serological result however can be equally important in excluding suspected tissue invasion by *E. histolytica*.

Prevention

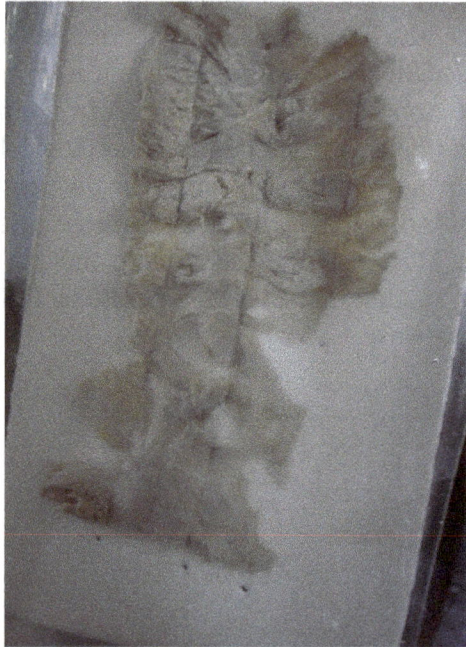

Amoebic ulcer in the intestines

To help prevent the spread of amoebiasis around the home :

- Wash hands thoroughly with soap and hot running water for at least 10 seconds after using the toilet or changing a baby's diaper, and before handling food.

- Clean bathrooms and toilets often; pay particular attention to toilet seats and taps.

- Avoid sharing towels or face washers.

To help prevent infection:

- Avoid raw vegetables when in endemic areas, as they may have been fertilized using human feces.

- Boil water or treat with iodine tablets.

- Avoid eating street foods especially in public places where others are sharing sauces in one container

Good sanitary practice, as well as responsible sewage disposal or treatment, are necessary for the prevention of *E. histolytica* infection on an endemic level. *E.histolytica* cysts are usually resistant to chlorination, therefore sedimentation and filtration of water supplies are necessary to reduce the incidence of infection.

E. histolytica cysts may be recovered from contaminated food by methods similar to those used for recovering *Giardia lamblia* cysts from feces. Filtration is probably the most practical method for recovery from drinking water and liquid foods. *E. histolytica* cysts must be distinguished from cysts of other parasitic (but nonpathogenic) protozoa and from cysts of free-living protozoa as discussed above. Recovery procedures are not very accurate; cysts are easily lost or damaged beyond recognition, which leads to many falsely negative results in recovery tests.

Treatment

E. histolytica infections occur in both the intestine and (in people with symptoms) in tissue of the intestine and/or liver. As a result, two different classes of drugs are needed to treat the infection, one for each location. Such anti-amoebic drugs are known as amoebicides.

Prognosis

In the majority of cases, amoebas remain in the gastrointestinal tract of the hosts. Severe ulceration of the gastrointestinal mucosal surfaces occurs in less than 16% of cases. In fewer cases, the parasite invades the soft tissues, most commonly the liver. Only rarely are masses formed (amoebomas) that lead to intestinal obstruction.(Mistaken for Ca caecum and appendicular mass) Other local complications include bloody diarrhea, pericolic and pericaecal abscess.

Complications of hepatic amoebiasis includes subdiaphragmatic abscess, perforation of diaphragm to pericardium and pleural cavity, perforation to abdominal cavital *(amoebic peritonitis)* and perforation of skin *(amoebiasis cutis)*.

Pulmonary amoebiasis can occur from hepatic lesion by haemotagenous spread and also by perforation of pleural cavity and lung. It can cause lung abscess, pulmono pleural fistula, empyema lung and broncho pleural fistula. It can also reach the brain through blood vessels and cause amoebic brain abscess and amoebic meningoencephalitis. Cutaneous amoebiasis can also occur in skin around sites of colostomy wound, perianal region, region overlying visceral lesion and at the site of drainage of liver abscess.

Urogenital tract amoebiasis derived from intestinal lesion can cause amoebic vulvovaginitis *(May's disease)*, rectovesicle fistula and rectovaginal fistula.

Entamoeba histolytica infection is associated with malnutrition and stunting of growth.

Epidemiology

As of 2010 it caused about 55,000 deaths down from 68,000 in 1990. In older textbooks it is often stated that 10% of the world's population is infected with *Entamoeba histolytica*. It is now known that at least 90% of these infections are due to *E. dispar*. Nevertheless, this means that there are up to 50 million true *E. histolytica* infections and approximately seventy thousand die each year, mostly from liver abscesses or other complications. Although usually considered a tropical parasite, the first case reported (in 1875) was actually in St Petersburg in Russia, near the Arctic Circle. Infection is more common in warmer areas, but this is both because of poorer hygiene and the parasitic cysts surviving longer in warm moist conditions.

History

The most dramatic incident in the USA was the Chicago World's Fair outbreak in 1933 caused by contaminated drinking water; defective plumbing permitted sewage to contaminate water. There were 1,000 cases (with 58 deaths). In 1998 there was an outbreak of amoebiasis in the Republic of Georgia. Between 26 May and 3 September 1998, 177 cases were reported, including 71 cases of intestinal amoebiasis and 106 probable cases of liver abscess.

Cryptosporidiosis

Cryptosporidiosis, also known as crypto, is a parasitic disease caused by *Cryptosporidium*, a genus of protozoan parasites in the phylum Apicomplexa. It affects the distal small intestine and can affect the respiratory tract in both immunocompetent (i.e., individuals with a normal functioning immune system) and immunocompromised (e.g., persons with HIV/AIDS) individuals, resulting in watery diarrhea with or without an unexplained cough. In immunocompromised individuals, the symptoms are particularly severe and can be fatal. It is primarily spread through the fecal-oral route, often through contaminated water; recent evidence suggests that it can also be transmitted via fomites in respiratory secretions.

Cryptosporidium is the organism most commonly isolated in HIV-positive patients presenting with diarrhea. Despite not being identified until 1976, it is one of the most common waterborne diseases and is found worldwide. The parasite is transmitted by environmentally hardy microbial cysts (oocysts) that, once ingested, exist in the small intestine and result in an infection of intestinal epithelial tissue.

Signs and Symptoms

Cryptosporidiosis may occur as an asymptomatic infection, an acute infection (i.e., duration shorter than 2 weeks), recurrent acute infections in which symptoms reappear following a brief period of recovery for up to 30 days, and a chronic infection (i.e., duration longer than 2 weeks) in which symptoms are severe and persistent. It may be fatal in individuals with a severely compromised immune system. Symptoms usually appear 5–10 days after infection (range: 2–28 days) and normally last for up to 2 weeks in immunocompetent individuals (i.e., individuals with a normal functioning immune system); symptoms are usually more severe and persist longer in immuno-compromised individuals (e.g., persons with HIV/AIDS). Following the resolution of diarrhea, symptoms can reoccur after several days or weeks due to reinfection. Based upon one clinical trial, the likelihood of re-infection is high in immunocompetent adults.

In immunocompetent individuals, cryptosporidiosis is primarily localized to the distal small intestine and sometimes the respiratory tract as well. In immunocompromised persons, cryptosporidiosis may disseminate to other organs, including the hepatobiliary system, pancreas, upper gastrointestinal tract, and urinary bladder; pancreatic and biliary infection can involve acalculous cholecystitis, sclerosing cholangitis, papillary stenosis, or pancreatitis.

Intestinal Cryptosporidiosis

Common signs and symptoms of intestinal cryptosporidiosis include:

- Moderate to severe watery diarrhea, sometimes contains mucus and rarely contains blood or leukocytes
    - In very severe cases, diarrhea may be profuse and cholera-like with malabsorption and hypovolemia
- Low-grade fever
- Crampy abdominal pain
- Dehydration
- Weight loss
- Fatigue
- Nausea and vomiting – suggests upper GI tract involvement and may lead to respiratory cryptosporidiosis
- Epigastric or right upper quadrant tenderness

Less common or rare signs and symptoms include:

- Reactive arthritis (may affect the hands, knees, ankles, and feet)
- Jaundice – suggests hepatobiliary involvement
- Ascites – suggests pancreatic involvement

Respiratory Cryptosporidiosis

Symptoms of upper respiratory cryptosporidiosis include:

- Inflammation of the nasal mucosa, sinuses, larynx, or trachea
- Nasal discharge
- Voice change (e.g., hoarseness)

Symptoms of lower respiratory cryptosporidiosis include:

- Cough

- Shortness of breath

- Fever

- Hypoxemia

Cause

Cryptosporidium is a genus of protozoan pathogens which is categorized under the phylum Apicomplexa. Other apicomplexan pathogens include the malaria parasite *Plasmodium*, and *Toxoplasma*, the causative agent of toxoplasmosis. A number of *Cryptosporidium* infect mammals. In humans, the main causes of disease are *C. parvum* and *C. hominis* (previously *C. parvum* genotype 1). *C. canis*, *C. felis*, *C. meleagridis*, and *C. muris* can also cause disease in humans. *Cryptosporidium* is capable of completing its life cycle within a single host, resulting in microbial cyst stages that are excreted in feces and are capable of transmission to a new host via the fecal-oral route. Other vectors of disease transmission also exist.

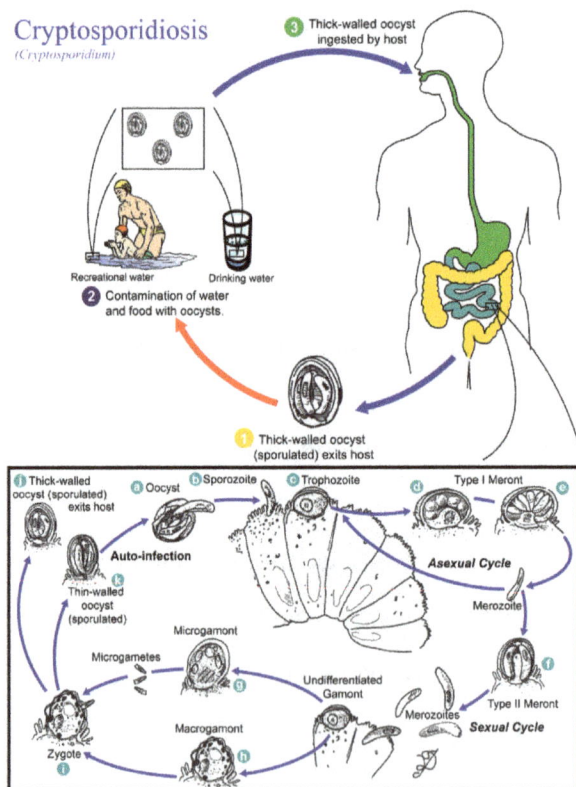

Life cycle of *Cryptosporidium* spp.

The pattern of *Cryptosporidium* life cycle fits well with that of other intestinal homogeneous coccidian genera of the suborder *Eimeriina*: macro- and microgamonts develop independently; a microgamont gives rise to numerous male gametes; and oocysts serving for parasites' spreading in the environment. Electron microscopic studies made from the 1970s have shown the intracellular, although extracytoplasmic localization of *Cryptosporidium* species.

These species possess a number of unusual features:

- an endogenous phase of development in microvilli of epithelial surfaces

- two morphofunctional types of oocysts

- the smallest number of sporozoites per oocyst

- a multi-membraneous "feeder" organelle

DNA studies suggest a relationship with the gregarines rather than the coccidia. The taxonomic position of this group has not yet been finally agreed upon.

The genome of *Cryptosporidium parvum* was sequenced in 2004 and was found to be unusual amongst Eukaryotes in that the mitochondria seem not to contain DNA. A closely related species, *C. hominis*, also has its genome sequence available. CryptoDB.org is a NIH-funded database that provides access to the *Cryptosporidium* genomics data sets.

Transmission

Infection is through contaminated material such as earth, water, uncooked or cross-contaminated food that has been in contact with the feces of an infected individual or animal. Contact must then be transferred to the mouth and swallowed. It is especially prevalent amongst those in regular contact with bodies of fresh water including recreational water such as swimming pools. Other potential sources include insufficiently treated water supplies, contaminated food, or exposure to feces. The high resistance of *Cryptosporidium* oocysts to disinfectants such as chlorine bleach enables them to survive for long periods and still remain infective. Some outbreaks have happened in day care related to diaper changes.

The following groups have an elevated risk of being exposed to *Cryptosporidium*:

- Child care workers

- Parents of infected children

- People who take care of other people with cryptosporidiosis

- International travelers

- Backpackers, hikers, and campers who drink unfiltered, untreated water

- People, including swimmers, who swallow water from contaminated sources

- People who handle infected cattle

- People exposed to human feces through sexual contact

Cases of cryptosporidiosis can occur in a city that does not have a contaminated water supply. In a city with clean water, it may be that cases of cryptosporidiosis have different origins. Testing of water, as well as epidemiological study, are necessary to determine the sources of specific infections. *Cryptosporidium* is causing serious illness more frequently in immunocompromised than in apparently healthy individuals. It may chronically sicken some children, as well as adults who are exposed and immunocompromised. A subset of the immunocompromised population is people with AIDS. Some sexual behaviours can transmit the parasite directly.

Life Cycle

Cryptosporidium spp. exist as multiple cell types which correspond to different stages in an infection (e.g., a sexual and asexual stage). As an oocyst – a type of hardy, thick-walled spore – it can survive in the environment for months and is resistant to many common disinfectants, particularly chlorine-based disinfectants. After being ingested, the oocysts exist in the small intestine. They release sporozoites that attach to the microvilli of the epithelial cells of the small intestine. From there they become trophozoites that reproduce asexually by multiple fission, a process known as schizogony. The trophozoites develop into Type 1 meronts that contain 8 daughter cells.

These daughter cells are Type 1 merozoites, which get released by the meronts. Some of these merozoites can cause autoinfection by attaching to epithelial cells. Others of these merozoites become Type II meronts, which contain 4 Type II merozoites. These merozoites get released and they attach to the epithelial cells. From there they become either macrogamonts or microgamonts. These are the female and male sexual forms, respectively. This stage, when sexual forms arise, is called gametogony.

Zygotes are formed by microgametes from the microgamont penetrating the macrogamonts. The zygotes develop into oocysts of two types. 20% of oocysts have thin walls and so can reinfect the host by rupturing and releasing sporozoites that start the process over again. The thick-walled oocysts are excreted into the environment. The oocysts are mature and infective upon being excreted.

Pathogenesis

The oocysts are ovoid or spherical and measure 5 to 6 micrometers across. When in flotation preparations they appear highly refractile. The oocysts contains up to 4 sporozoites that are bow-shaped.

As few as 2 to 10 oocysts can initiate an infection. The parasite is located in the brush border of the epithelial cells of the small intestine. They are mainly located in the jejunum. When the sporozoites attach the epithelial cells' membrane envelops them. Thus, they are "intracellular but extracytoplasmic". The parasite can cause damage to the microvilli where it attaches. The infected human excretes the most oocysts during the first week. Oocysts can be excreted for weeks after the diarrhea subsides from infections by *C. parvum* or *C. hominis*; however, immunocompetent individuals with *C. muris* infections have been observed excreting oocysts for seven months.

The immune system reduces the formation of Type 1 merozoites as well as the number of thinwalled oocysts. This helps prevent autoinfection. B cells do not help with the initial response or the fight to eliminate the parasite.

Diagnosis

There are many diagnostic tests for *Cryptosporidium*. They include microscopy, staining, and detection of antibodies. Microscopy can help identify oocysts in fecal matter. To increase the chance of finding the oocysts, the diagnostician should inspect at least 3 stool samples. There are several techniques to concentrate either the stool sample or the oocysts. The modified formalin-ethyl acetate (FEA) concentration method concentrates the stool. Both the modified zinc sulfate centrifugal flotation technique and the Sheather's sugar flotation procedure can concentrate the oocysts by causing them to float. Another form of microscopy is fluorescent microscopy done by staining with auramine.

Other staining techniques include acid-fast staining, which will stain the oocysts red. One type of acid-fast stain is the Kinyoun stain. Giemsa staining can also be performed. Part of the small intestine can be stained with hematoxylin and eosin (H & E), which will show oocysts attached to the epithelial cells.

Detecting antigens is yet another way to diagnose the disease. This can be done with direct fluorescent antibody (DFA) techniques. It can also be achieved through indirect immunofluorescence assay. Enzyme-linked immunosorbent assay (ELISA) also detects antigens.

Polymerase chain reaction (PCR) is another way to diagnose cryptosporidiosis. It can even identify the specific species of *Cryptosporidium*. If the patient is thought to have biliary cryptosporidiosis, then an appropriate diagnostic technique is ultrasonography. If that returns normal results, the next step would be to perform endoscopic retrograde cholangiopancreatography.

Prevention

Many treatment plants that take raw water from rivers, lakes, and reservoirs for public drinking water production use conventional filtration technologies. This involves a series of processes, including coagulation, flocculation, sedimentation, and filtration. Direct filtration, which is typically used to treat water with low particulate levels, includes coagulation and filtration, but not sedimentation. Other common filtration processes, including slow sand filters, diatomaceous earth filters and membranes will remove 99% of *Cryptosporidium*. Membranes and bag and cartridge filters remove *Cryptosporidium* product-specifically.

While *Cryptosporidium* is highly resistant to chlorine disinfection, with high enough concentrations and contact time, *Cryptosporidium* will be inactivated by chlorine dioxide and ozone treatment. The required levels of chlorine generally preclude the use of chlorine disinfection as a reliable method to control *Cryptosporidium* in drinking water. Ultraviolet light treatment at relatively low doses will inactivate *Cryptosporidium*. Water Research Foundation-funded research originally discovered UV's efficacy in inactivating *Cryptosporidium*.

One of the largest challenges in identifying outbreaks is the ability to identify *Cryptosporidium* in the laboratory. Real-time monitoring technology is now able to detect *Cryptosporidium* with online systems, unlike the spot and batch testing methods used in the past.

The most reliable way to decontaminate drinking water that may be contaminated by *Cryptosporidium* is to boil it.

In the US the law requires doctors and labs to report cases of cryptosporidiosis to local or state health departments. These departments then report to the Center for Disease Control and Prevention. The best way to prevent getting and spreading cryptosporidiosis is to have good hygiene and sanitation. An example would be hand-washing. Prevention is through washing hands carefully after going to the bathroom or contacting stool, and before eating. People should avoid contact with animal feces. They should also avoid possibly contaminated food and water. In addition, people should refrain from engaging in sexual activities that can expose them to feces.

Standard water filtration may not be enough to eliminate *Cryptosporidium*; boiling for at least 1 minute (3 minutes above 6,500 feet (2,000 m) of altitude) will decontaminate it. Heating milk at 71.7 °C (161 °F) for 15 seconds pasteurizes it and can destroy the oocysts' ability to infect. Water can also be made safe by filtering with a filter with pore size not greater than 1 micrometre, or by filters that have been approved for "cyst removal" by NSF International National Sanitation Foundation. Bottled drinking water is less likely to contain *Cryptosporidium*, especially if the water is from an underground source.

People with cryptosporidiosis should not swim in communal areas because the pathogen can reside in the anal and genital areas and be washed off. They should wait until at least two weeks after diarrhea stops before entering public water sources, since oocysts can still be shed for a while. Also, they should stay away from immunosuppressed people. Immunocompromised people should take care to protect themselves from water in lakes and streams. They should also stay away from animal stools and wash their hands after touching animals. To be safe, they should boil or filter their water. They should also wash and cook their vegetables.

The US CDC notes the recommendation of many public health departments to soak contaminated surfaces for 20 minutes with a 3% hydrogen peroxide (99% kill rate) and then rinse them thoroughly, with the caveat that no disinfectant is guaranteed to be completely effective against Cryptosporidium. However, hydrogen peroxide is more effective than standard bleach solutions.

Treatment

Symptomatic treatment primarily involves fluid rehydration, electrolyte replacement (sodium, potassium, bicarbonate, and glucose), and antimotility agents (e.g., loperamide). Supplemental zinc may improve symptoms, particularly in recurrent or persistent infections or in others at risk for zinc deficiency.

Immunocompetent

Immunocompetent individuals with cryptosporidiosis typically suffer a short (i.e., duration of less than 2 weeks) self-limiting course of diarrhea that may require symptomatic treatment and ends with spontaneous recovery; in some circumstances, antiparasitic medication may be required (e.g., recurrent, severe, or persistent symptoms); however reinfection frequently occurs.

As of 2015, nitazoxanide is the only antiparasitic drug treatment with proven efficacy for cryptosporidiosis in immunocompetent individuals; however, it lacks efficacy in severely immunocompromised patients. Certain agents such as paromomycin and azithromycin are sometimes used as well, but they only have partial efficacy.

Immunocompromised

In immunocompromised individuals, such as AIDS patients, cryptosporidiosis resolves slowly or not at all, and frequently causes a particularly severe and persistent form of watery diarrhea coupled with a greatly decreased ability to absorb key nutrients through the intestinal tract. As a result, infected individuals may experience severe dehydration, electrolyte imbalances, malnutrition, wasting, and potentially death. In general, the mortality rate for infected AIDS patients is based on CD4+ marker counts. Patients with CD4+ counts over 180 cells/mm³ recover with supportive hospital care and medication; but, in patients with CD4+ counts below 50 cells/mm³, the effects are usually fatal within 3 to 6 months. During the Milwaukee cryptosporidiosis epidemic (the largest of its kind), 73% of AIDS patients with CD4+ counts lower than 50 cells/mm³ and 36% of those with counts between 50 and 200 cells/mm³ died within the first year of contracting the infection.

The best treatment approach is to improve the immune status in immunodeficient individuals using highly active antiretroviral therapy that includes an HIV protease inhibitor along with continued use of antiparasitic medication. Antiparasitic drug treatment for immunocompromised individuals usually involves the combination of nitazoxanide, paromomycin, and azithromycin together; these drugs are only partially active in HIV/AIDS patients compared to their effect in immunocompetent persons. A Cochrane Collaboration review recommended that nitazoxanide be considered for use in treatment despite its reduced effectiveness in immunocompromised individuals.

Currently, research is being done in molecular-based immunotherapy. For example, synthetic isoflavone derivates have been shown to fight off *Cryptosporidium parvum* both *in vitro* and in animal studies. Derivates of nitazoxanide, known as thiazolides, have also shown promising results *in vitro*.

Epidemiology

Cryptosporidiosis is found worldwide. It causes 50.8% of water-borne diseases that are attributed to parasites. In developing countries, 8–19% of diarrheal diseases can be attributed to *Cryptosporidium*. Ten percent of the population in developing countries excretes oocysts. In developed countries, the number is lower at 1–3%. The age group most affected are children from 1 to 9 years old.

Roughly 30% of adults in the United States are seropositive for cryptosporidiosis, meaning that they contracted the infection at some point in their lives.

History

The organism was first described in 1907 by Tyzzer, who recognised it was a coccidian.

Research

A recombinant *Cryptosporidium parvum* oocyst surface protein (rCP15/60) vaccine has produced an antibody response in a large group of cows and also antibody response in calves fed rCP15/60-immune colostrum produced by these vaccinated cows . This is very promising. Human *Cryptosporidium parvum* infections are particularly prevalent and often fatal in neonates in developing countries and to immunocompromised people, such as AIDs patients. To date (2013), there is no commercially available effective vaccine against *Cryptosporidium parvum*, although passive immunization utilizing different zoite surface (glyco)proteins has showed promise. Developmental stages within the life cycle of the parasite that might act as possible targets for vaccine development. The organism is detected in 65–97% of the surface-water supply in the United States and is resistant to most disinfectants used for treatment of drinking water. Antibodies in the serum of humans and animals infected with *Cryptosporidium parvum* react with several antigens, one of which is a 15 kDa protein (CP15) located on the surface of the organism. This protein is a good candidate for use as a molecular vaccine because previous studies have shown that a monoclonal antibody to CP15 confers passive immunity to mice. Currently, there is no vaccine or completely effective drug therapy against *Cryptosporidium parvum* in HIV/AIDS individuals.

Other Animals

The most important zoonotic reservoirs are cattle, sheep and goats. In addition, in recent years, cryptosporidiosis has plagued many commercial leopard gecko breeders. Several species of the Cryptosporidium family (C. serpentes and others) are involved, and outside of geckos it has been found in monitor lizards, iguanas and tortoises, as well as several snake species.

Notable cases

Before 2000

- In 1987, 13,000 people in Carrollton, Georgia, United States, became ill with cryptosporidiosis. This was the first report of its spread through a municipal water system that met all state and federal drinking water standards.

- In 1993, a waterborne cryptosporidiosis outbreak occurred in Milwaukee, Wisconsin, US. An estimated 403,000 people became ill, including 4,400 people hospitalized. An estimated 69 people died during the outbreak, according to the CDC.

- The UK's biggest outbreak occurred in Torbay in Devon in 1995.

- In the summer of 1996, *Cryptosporidium* affected approximately 2,000 people in Cranbrook, British Columbia, Canada. Weeks later, a separate incident occurred in Kelowna, British Columbia, where 10,000 to 15,000 people got sick.

2001–2009

- In April 2001, an outbreak occurred in the city of North Battleford, Saskatchewan, Canada. Between 5800 and 7100 people suffered from diarrheal illness, and 1907 cases of cryptosporidiosis were confirmed. Equipment failures at the city's antiquated water filtration plant following maintenance were found to have caused the outbreak.

- In the summer of 2005, after numerous reports by patrons of gastrointestinal upset, a water park at Seneca Lake State Park, in the Finger Lakes region of upstate New York was found to have two water storage tanks infected with *Cryptosporidium*. By early September 2005, over 3,800 people reported symptoms of a *Cryptosporidium* infection. The "Sprayground" was ordered closed for the season on 15 August.

- In October 2005, the Gwynedd and Anglesey areas of North Wales, United Kingdom, suffered an outbreak of cryptosporidiosis. The outbreak may have been linked to the drinking water supply from Llyn Cwellyn, but this is not yet confirmed. As a result, 231 people fell ill and the company Welsh Water (Dwr Cymru) advised 61,000 people to boil their water before use.

- In March 2007, a suspected outbreak occurred in Galway, Ireland, after the source of water for much of the county, Lough Corrib, was suspected to be contaminated with the parasite. A large population (90,000 people), including areas of both Galway City and County, were advised to boil water for drinking, food preparation and for brushing teeth. On 21 March 2007, it was confirmed that the city and county's water supply was contaminated with the parasite. The area's water supply was finally given approval on 20 August 2007, five months after *Cryptosporidium* was first detected. Around 240 people are known to have contracted the disease; experts say the true figure could be up to 5,000.

- Hundreds of public pools in 20 Utah counties were closed to young children in 2007, as children under 5 are most likely to spread the disease, especially children wearing diapers. As of 10 September 2007 the Utah Department of Health had reported 1302 cases of cryptosporidiosis in the year; a more usual number would be 30. On 25 September the pools were reopened to those not requiring diapers, but hyperchlorination requirements were not lifted.

- On 21 September 2007, a *Cryptosporidium* outbreak attacked the Western United States: 230 Idaho residents, with hundreds across the Rocky Mountain area; in the Boise and Meridian areas; Utah, 1,600 illnesses; Colorado and other Western states — Montana, decrease.

- On 25 June 2008, *Cryptosporidium* was found in England in water supplies in Northampton, Daventry, and some surrounding areas supplied from the Pitsford Reservoir, as reported on the BBC. People in the affected areas were warned not to drink tap water unless it had been boiled. Anglian Water confirmed that 108,000 households were affected, about 250,000 people. They advised that water might not be fit for human consumption for many weeks. The boil notice was lifted for all the affected customers on 4 July 2008.

- Throughout the summer of 2008; many public swimming areas, water parks, and public pools in the Dallas/Fort Worth Metroplex of Texas suffered an outbreak of cryptosporidiosis. Burger's Lake in Fort Worth was the first to report such an outbreak. This prompted some, if not all, city-owned and private pools to close and hyperchlorinate. To the 13 August 2008 there were 400 reported cases of *Cryptosporidium*.

- In September 2008, a gym in Cambridge, United Kingdom, was forced to close its swimming pool until further notice after health inspectors found an outbreak of cryptosporidiosis. Environmental Health authorities requested that the water be tested after it was confirmed that a young man had been infected.

2010 and Later

- In May 2010, the Behana creek water supply south of Cairns, Australia, was found to be contaminated by cryptosporidium.

- In July 2010, a local sport centre in Cumbernauld (Glasgow, UK) detected traces of cryptosporidium in its swimming pools, causing a temporarily closure of the swimming pools.

- In November 2010, over 4000 cases of cryptosporidiosis were reported in Östersund, Sweden. The source of contamination was the tap water. In mid December 2010 the number of reported cases was 12,400 according to local media.

- As of April 2011, there has been an ongoing outbreak in Skellefteå, Sweden. Although many people have been diagnosed with cryptosporidiosis, the source of the parasite has not yet been found. Several tests have been taken around the water treatment unit "Abborren", but so far no results have turned up positive. Residents are being advised to boil the tap water as they continue to search for the contaminating source.

- Since May 2011, there has been an ongoing outbreak in South Roscommon in Ireland. Although many people have been diagnosed with cryptosporidiosis, the source of the parasite has not yet been found. Testing continues and Roscommon County Council are now considering introducing Ultra Violet Filtration to their water treatment process in the next 12 months. Residents are being advised to boil the tap water and there is no sign of this boil notice being lifted in the near future.

- In May 2013, in Roscommon, Ireland, another outbreak of the cryptosporidiosis was reported and a boil water notice was issued. This was the second time the parasite was detected in a month in the Roscommon water supply. The source of one of the outbreaks had been linked to the agricultural community. To date, 13 people have been treated for Cryptosporidiosis and the boil water notice is still in effect.

Schistosomiasis

Schistosomiasis, also known as snail fever, is a disease caused by parasitic flatworms called schistosomes. The urinary tract or the intestines may be infected. Signs and symptoms may include abdominal pain, diarrhea, bloody stool, or blood in the urine. Those who have been infected a long time may experience liver damage, kidney failure, infertility, or bladder cancer (squamous cell carcinoma). In children, it may cause poor growth and learning difficulty.

The disease is spread by contact with fresh water contaminated with the parasites. These parasites are released from infected freshwater snails. The disease is especially common among children in developing countries as they are more likely to play in contaminated water. Other high risk groups include farmers, fishermen, and people using unclean water during daily living. It belongs to the group of helminth infections. Diagnosis is by finding eggs of the parasite in a person's urine or stool. It can also be confirmed by finding antibodies against the disease in the blood.

Methods to prevent the disease include improving access to clean water and reducing the number of snails. In areas where the disease is common, the medication praziquantel may be given once a year to the entire group. This is done to decrease the number of people infected and, consequently, the spread of the disease. Praziquantel is also the treatment recommended by the World Health Organization (WHO) for those who are known to be infected.

Schistosomiasis affected almost 210 million people worldwide as of 2012. An estimated 12,000 to 200,000 people die from it each year. The disease is most commonly found in Africa, as well as Asia and South America. Around 700 million people, in more than 70 countries, live in areas where the disease is common. In tropical countries, schistosomiasis is second only to malaria among parasitic diseases with the greatest economic impact. Schistosomiasis is listed as a neglected tropical disease.

Signs and Symptoms

Many individuals do not experience symptoms. If symptoms do appear, it usually takes four to six weeks from the time of infection. The first symptom of the disease may be a general ill feeling. Within twelve hours of infection, an individual may complain of a tingling sensation or light rash, commonly referred to as "swimmer's itch", due to irritation at the point of entrance. The rash that may develop can mimic scabies and other types of rashes. Other symptoms can occur two to ten weeks later and can include fever, aching, cough, diarrhea, or gland enlargement. These symptoms can also be related to avian schistosomiasis, which does not cause any further symptoms in humans.

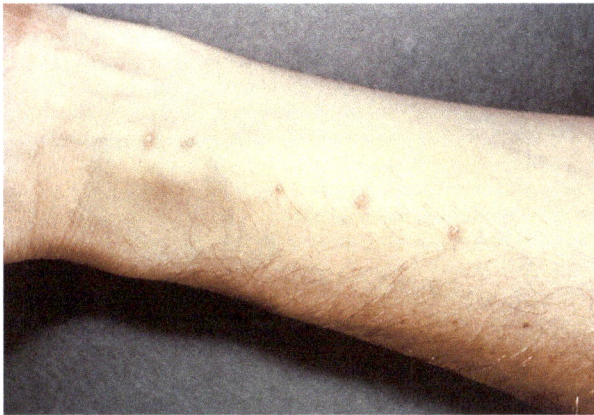

Skin blisters on the forearm, created by the entrance of *Schistosoma* parasite

The manifestations of schistosomal infection vary over time as the cercariae, and later adult worms and their eggs migrate through the body.

Intestinal Schistosomiasis

In intestinal schistosomiasis, eggs become lodged in the intestinal wall and cause an immune system reaction called a granulomatous reaction. This immune response can lead to obstruction of the colon and blood loss. The infected individual may have what appears to be a potbelly. Eggs can also become lodged in the liver, leading to high blood pressure through the liver, enlarged spleen, the buildup of fluid in the abdomen, and potentially life-threatening dilations or swollen areas in the esophagus or gastrointestinal tract that can tear and bleed profusely (esophageal varices). In rare instances, the central nervous system is affected. Individuals with chronic active schistosomiasis may not complain of typical symptoms.

Dermatitis

The first potential reaction is an itchy, papular rash that results from cercariae penetrating the skin, often in a person's first infection. The round bumps are usually one to three centimeters big.

Because people living in affected areas have often been repeatedly exposed, acute reactions are more common in tourists and migrants. The rash can occur between the first few hours and a week after exposure and lasts for several days. A similar, more severe reaction called "swimmer's itch" reaction can also be caused by cercariae from animal trematodes that often infect birds.

Katayama Fever

Another primary condition, called Katayama fever, may also develop from infection with these worms, and it can be very difficult to recognize. Symptoms include fever, lethargy, the eruption of pale temporary bumps associated with severe itching (urticarial) rash, liver and spleen enlargement, and bronchospasm.

Acute schistosomiasis (Katayama fever) may occur weeks or months after the initial infection as a systemic reaction against migrating schistosomulae as they pass through the bloodstream through the lungs to the liver. Similarly to swimmer's itch, Katayama fever is more commonly seen in people with their first infection such as migrants and tourists. However it is seen in native residents of China infected with *S. japonicum*. Symptoms include:

- Dry cough with changes on chest x-ray
- Fever
- Fatigue
- Muscle aches
- Malaise
- Abdominal pain
- Enlargement of both the liver and the spleen

The symptoms usually get better on their own but a small proportion of people have persistent weight loss, diarrhea, diffuse abdominal pain and rash.

Chronic Disease

In long established disease adult worms lay eggs that can cause inflammatory reactions. The eggs secrete proteolytic enzymes that help them migrate to the bladder and intestines to be shed. However the enzymes also cause an eosinophilic inflammatory reaction when eggs get trapped in tissues or embolize to the liver, spleen, lungs or brain. The long term manifestations are dependent on the species of schistosome as the adult worms of different species migrate to different areas. Many infections are mildly symptomatic, with anemia and malnutrition being common in endemic areas.

Genitourinary Disease

The worms of *S. haematobium* migrate to the veins around the bladder and ureters. This can lead to blood in the urine 10 to 12 weeks after infection. Over time, fibrosis can lead to obstruction of the urinary tract, hydronephrosis and kidney failure. Bladder cancer diagnosis and mortality are generally elevated in affected areas, and have efforts to control schistosomiasis in Egypt have led to decreases in the bladder cancer rate. The risk of bladder cancer appears to be especially high in male smokers, perhaps due to chronic irritation of the bladder lining allowing it to be exposed to carcinogens from smoking.

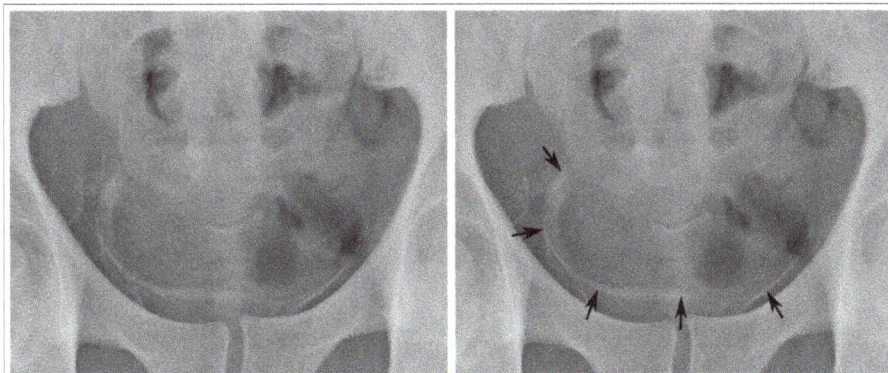

Calcification of the bladder wall on a plain x-ray image of the pelvis, in a 44-year-old sub-Saharan man. This is due to urinary schistosomiasis.

In women, genitourinary disease can also include genital lesions that may lead to increased rates of HIV transmission.

Gastrointestinal Disease

The worms of *S. mansoni* and *S. japonicum* migrate to the veins of the gastrointestinal tract and liver. Eggs in the gut wall can lead to pain, blood in the stool, and diarrhea (especially in children). Severe disease can lead to narrowing of the colon or rectum. Eggs also migrate to the liver leading to fibrosis in 4 to 8 percent of people with chronic infection, mainly those with long term heavy infection.

Central Nervous System Disease

Central nervous system lesions occur occasionally. Cerebral granulomatous disease may be caused by *S. japonicum* eggs in the brain. Communities in China affected by S. japonicum have been found to have rates of seizures eight times higher than baseline. Similarly, granulomatous lesions from S. mansoni and S. haematobium eggs in the spinal cord can lead to transverse myelitis with flaccid paraplegia. Eggs are thought to travel to the central nervous system via embolization.

Diagnosis

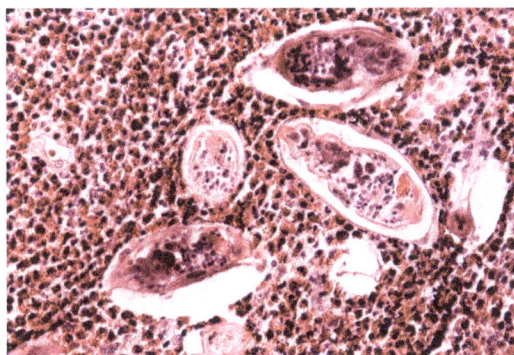

High powered detailed micrograph of *Schistosoma* parasite eggs in human bladder tissue.

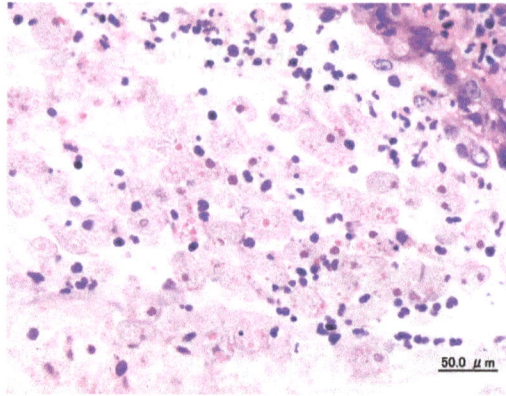

S. japonicum eggs in hepatic portal tract.

Identification of Eggs in Stools

Diagnosis of infection is confirmed by the identification of eggs in stools. Eggs of *S. mansoni* are approximately 140 by 60 μm in size, and have a lateral spine. The diagnosis is improved by the use of the Kato-Katz technique (a semi-quantitative stool examination technique). Other methods that can be used are enzyme-linked immunosorbent assay (ELISA), circumoval precipitation test (COPT), and alkaline phosphatase immunoassay (APIA).

Microscopic identification of eggs in stool or urine is the most practical method for diagnosis. Stool examination should be performed when infection with *S. mansoni* or *S. japonicum* is suspected, and urine examination should be performed if *S. haematobium* is suspected. Eggs can be present in the stool in infections with all *Schistosoma* species. The examination can be performed on a simple smear (1 to 2 mg of fecal material). Since eggs may be passed intermittently or in small amounts, their detection will be enhanced by repeated examinations and/or concentration procedures. In addition, for field surveys and investigational purposes, the egg output can be quantified by using the Kato-Katz technique (20 to 50 mg of fecal material) or the Ritchie technique. Eggs can be found in the urine in infections with *S. haematobium* (recommended time for collection: between noon and 3 PM) and with *S. japonicum*. Quantification is possible by using filtration through a nucleopore filter membrane of a standard volume of urine followed by egg counts on the membrane. Tissue biopsy (rectal biopsy for all species and biopsy of the bladder for *S. haematobium*) may demonstrate eggs when stool or urine examinations are negative.

Antibody Detection

Antibody detection can be useful to indicate schistosome infection in people who have traveled to areas were schistosomiasis is common and in whom eggs cannot be demonstrated in fecal or urine specimens. Test sensitivity and specificity vary widely among the many tests reported for the serologic diagnosis of schistosomiasis and are dependent on both the type of antigen preparations used (crude, purified, adult worm, egg, cercarial) and the test procedure.

At CDC, a combination of tests with purified adult worm antigens are used for antibody detection. All serum specimens are tested by FAST-ELISA using *S. mansoni* adult microsomal antigen (MAMA). A positive reaction (greater than 9 units/μl serum) indicates infection with *Schisto-*

soma species. Sensitivity for *S. mansoni* infection is 99 percent, 95 percent for *S. haematobium* infection, and less than 50 percent for *S. japonicuminfection*. Specificity of this assay for detecting schistosome infection is 99 percent. Because test sensitivity with the FAST-ELISA is reduced for species other than *S. mansoni*, immunoblots of the species appropriate to the patient's travel history are also tested to ensure detection of *S. haematobium* and *S. japonicum* infections. Immunoblots with adult worm microsomal antigens are species-specific and so a positive reaction indicates the infecting species. The presence of antibody is indicative only of schistosome infection at some time and cannot be correlated with clinical status, worm burden, egg production, or prognosis. Where a person has traveled can help determine what *Schistosoma* species to test for by immunoblot.

In 2005 a field evaluation of a novel handheld microscope was undertaken in Uganda for the diagnosis of intestinal schistosomiasis by a team led by Russell Stothard from the Natural History Museum of London, working with the Schistosomiasis Control Initiative, London.

Prevention

Many countries are working towards eradicating the disease. The WHO is promoting these efforts. In some cases, urbanization, pollution, and consequent destruction of snail habitat have reduced exposure, with a subsequent decrease in new infections. Furthermore, the drug praziquantel is used for prevention in high-risk populations living in areas where the disease is common.

A 2014 review found tentative evidence that increasing access to clean water and sanitation reduces schistosome infection.

Snails

Prevention is best accomplished by eliminating the water-dwelling snails that are the natural reservoir of the disease.

For many years from the 1950s onwards, vast dams and irrigation schemes were constructed, causing a massive rise in water-borne infections from schistosomiasis. The detailed specifications laid out in various UN documents since the 1950s could have minimized this problem. Irrigation schemes can be designed to make it hard for the snails to colonize the water, and to reduce the contact with the local population.

Even though guidelines on how to design these schemes to minimise the spread of the disease had been published years before, the designers were unaware of them.

Treatment

Currently there are two drugs available, praziquantel and oxamniquine, for the treatment of schistosomiasis. They are considered equivalent in relation to efficacy against S. mansoni and safety. Due to its lower cost per treatment, and oxaminiquine's lack of efficacy against the urogenital form of the disease caused by S. haematobium, in general praziquantel is considered the first option for treatment. The treatment objective is to cure the disease and to prevent the evolution of the acute to the chronic form of the disease. All cases of suspected schistosomiasis should be treated regardless of presentation because the adult parasite can live in the host for years.

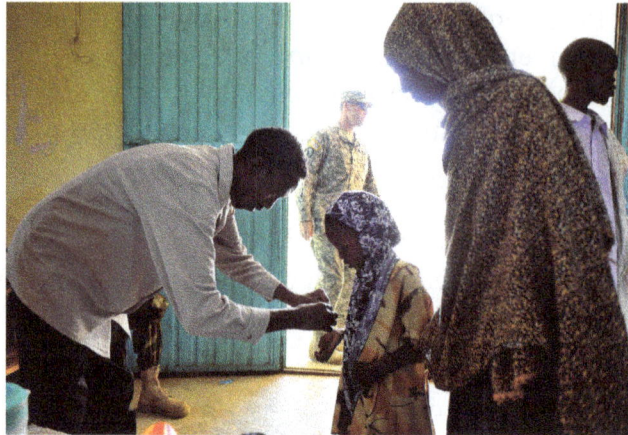

Ethiopian children treated for *Schistosoma mansoni*

Schistosomiasis is treatable by taking by mouth a single dose of the drug praziquantel annually.

The WHO has developed guidelines for community treatment based on the impact the disease has on children in villages in which it is common:

- When a village reports more than 50 percent of children have blood in their urine, everyone in the village receives treatment.

- When 20 to 50 percent of children have bloody urine, only school-age children are treated.

- When fewer than 20 percent of children have symptoms, mass treatment is not implemented.

Other possible treatments include a combination of praziquantel with metrifonate, artesunate, or mefloquine. A Cochrane review found tentative evidence that when used alone, metrifonate was as effective as praziquantel.

Another agent, mefloquine, which has previously been used to treat and prevent malaria, was recognised in 2008–2009 to be effective against *Schistosoma*.

Epidemiology

The disease is found in tropical countries in Africa, the Caribbean, eastern South America, Southeast Asia, and the Middle East. *S. mansoni* is found in parts of South America and the Caribbean, Africa, and the Middle East; *S. haematobium* in Africa and the Middle East; and *S. japonicum* in the Far East. *S. mekongi* and *S. intercalatum* are found locally in Southeast Asia and central West Africa, respectively.

The disease is endemic in about 75 developing countries and mainly affects people living in rural agricultural and peri-urban areas.

Infection Estimates

In 2010, approximately 238 million people were infected with schistosomiasis, 85 percent of whom live in Africa. An earlier estimate from 2006 had put the figure at 200 million people infected. In many of the affected areas, schistosomiasis infects a large proportion of children under 14 years of age. An estimated 600 to 700 million people worldwide are at risk from the disease because they live in countries where the organism is common. In 2012, 249 million people were in need of treatment to prevent the disease. This likely makes it the most common parasitic infection with malaria second and causing about 207 million cases in 2013.

S. haematobium, the infectious agent responsible for urogenital schistosomiasis, infects over 112 million people annually in Sub-Saharan Africa alone. It is responsible for 32 million cases of dysuria, 10 million cases of hydronephrosis, and 150,000 deaths from renal failure annually, making *S. haematobium* the world's deadliest schistosome.

Deaths

Estimates regarding the number of deaths vary. Worldwide in 2010 the Global Burden of Disease estimated 12,000 direct deaths while the WHO estimates more than 200,000 people die related to schistosomiasis yearly. Another 20 million have severe consequences from the disease. It is the most deadly of the neglected tropical diseases.

History

Schistosomiasis is known as bilharzia or bilharziosis in many countries, after German physician Theodor Bilharz, who first described the cause of urinary schistosomiasis in 1851.

The first physician who described the entire disease cycle was Brazilian parasitologist Pirajá da Silva in 1908. The first known case of infection was discovered in 2014, belonging to a child who lived 6,200 years ago.

It was a common cause of death for Egyptians in the Greco-Roman Period.

Society and Culture

Schistosomiasis is endemic in Egypt, exacerbated by the country's dam and irrigation projects along the Nile. From the late 1950s through the early 1980s, infected villagers were treated with repeated injections of tartar emetic. Epidemiological evidence suggests that this campaign unintentionally contributed to the spread of hepatitis C via unclean needles. Egypt has the world's highest hepatitis C infection rate, and the infection rates in various regions of the country closely track the timing and intensity of the anti-schistosomiasis campaign. From ancient times to the early 20th century, schistosomiasis' symptom of blood in the urine was seen as a male version of menstruation in Egypt and was thus viewed as a rite of passage for boys.

Schistosomiasis was mentioned in an episode of the television sitcom *WKRP in Cincinnati*. In the episode "Frog Story" from season 3, Dr. Johnny Fever becomes concerned that he may have schistosomiasis. Among human parasitic diseases, schistosomiasis ranks second behind malaria in terms of socio-economic and public health importance in tropical and subtropical areas.

Research

Vaccine

As with other major parasitic diseases, there is ongoing research into developing a schistosomiasis vaccine that will prevent the parasite from completing its life cycle in humans. As of September

2014 Eurogentec Biologics was developing a vaccine against bilharziosis called *Bilhvax* in partnership with INSERM and researchers from the Pasteur Institute; the vaccine candidate was starting Phase III trials at that time.

Dracunculiasis

Dracunculiasis, also called Guinea-worm disease (GWD), is an infection by the Guinea worm. A person becomes infected when they drink water that contains water fleas infected with guinea worm larvae. Initially there are no symptoms. About one year later, the person develops a painful burning feeling as the female worm forms a blister in the skin, usually on the lower limb. The worm then comes out of the skin over the course of a few weeks. During this time, it may be difficult to walk or work. It is very uncommon for the disease to cause death.

In humans, the only known cause is *Dracunculus medinensis*. The worm is about one to two millimeters wide, and an adult female is 60 to 100 centimeters long (males are much shorter at 12–29 mm or 0.47–1.14 in). Outside of humans, the young form can survive up to three weeks, during which they must be eaten by water fleas to continue to develop. The larva inside water fleas may survive up to four months. Thus, in order for the disease to remain in an area, it must occur each year in humans. A diagnosis of the disease can usually be made based on the signs and symptoms of the disease.

Prevention is by early diagnosis of the disease followed by keeping the person from putting the wound in drinking water to decrease spread of the parasite. Other efforts include improving access to clean water and otherwise filtering water if it is not clean. Filtering through a cloth is often enough. Contaminated drinking water may be treated with a chemical called temefos to kill the larva. There is no medication or vaccine against the disease. The worm may be slowly removed over a few weeks by rolling it over a stick. The ulcers formed by the emerging worm may get infected by bacteria. Pain may continue for months after the worm has been removed.

In 2015 there were 22 reported cases of the disease and in the first half of 2016 there were 2 confirmed cases. This is down from an estimated 3.5 million cases in 1986. It only exists in 4 countries in Africa, down from 20 countries in the 1980s. It will likely be the first parasitic disease to be globally eradicated. Guinea worm disease has been known since ancient times. It is mentioned in the Egyptian medical Ebers Papyrus, dating from 1550 BC. The name dracunculiasis is derived from the Latin "affliction with little dragons", while the name "guinea worm" appeared after Europeans saw the disease on the Guinea coast of West Africa in the 17th century. Other *Dracunculus* species are known to infect various mammals, but do not appear to infect humans. Dracunculiasis is classified as a neglected tropical disease. Because dogs may also become infected, the eradication program is monitoring and treating dogs as well.

Signs and Symptoms

Dracunculiasis is diagnosed by seeing the worms emerging from the lesions on the legs of infected individuals and by microscopic examinations of the larvae.

As the worm moves downwards, usually to the lower leg, through the subcutaneous tissues, it leads to intense pain localized to its path of travel. The painful, burning sensation experienced by infected people has led to the disease being called "the fiery serpent". Other symptoms include fever, nausea, and vomiting. Female worms cause allergic reactions during blister formation as they migrate to the skin, causing an intense burning pain. Such allergic reactions produce rashes, nausea, diarrhea, dizziness, and localized edema. When the blister bursts, allergic reactions subside, but skin ulcers form, through which the worm can protrude. Only when the worm is removed is healing complete. Death of adult worms in joints can lead to arthritis and paralysis in the spinal cord.

Cause

Dracunculiasis is caused by drinking water contaminated by water fleas that host the *D. medinensis* larvae. Dracunculiasis has a history of being very common in some of the world's poorest areas, particularly those with limited or no access to clean water. In these areas, stagnant water sources may still host copepods, which can carry the larvae of the guinea worm.

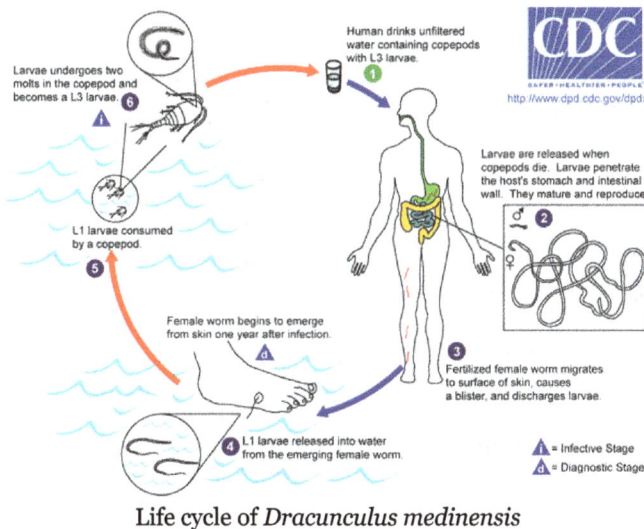

Life cycle of *Dracunculus medinensis*

Humans and dogs are the only known animals that guinea worms infect. Other species in the Dracunculus genus affect other mammals.

After ingestion, the copepods die and are digested, thus releasing the stage 3 larvae, which then penetrate the host's stomach or intestinal wall, and then enter into the abdominal cavity and retroperitoneal space. After maturation, which takes approximately three months, mating takes place; the male worm dies after mating and is absorbed by the host's body.

Approximately one year after mating, the fertilized females migrate in the subcutaneous tissues adjacent to long bones or joints of the extremities. They then move towards the surface, resulting in blisters on the skin, generally on the distal lower extremity (foot). Within 72 hours, the blister ruptures, exposing one end of the emergent worm. The blister causes a very painful burning sensation as the worm emerges, and the sufferer will often immerse the affected limb in water to relieve the burning sensation. When a blister or open sore is submerged in water, the adult female releases hundreds of thousands of stage 1 guinea worm larvae, thereby contaminating the water.

During the next few days, the female worm can release more larvae whenever it comes in contact with water, as it extends its posterior end through the hole in the host's skin. These larvae are eaten by copepods, and after two weeks (and two molts), the stage 3 larvae become infectious and, if not filtered from drinking water, will cause the cycle to repeat. Infected copepods can live in the water for up to 4 months.

The male guinea worm is typically much smaller (12–29 mm or 0.47–1.14 in) than the female, which, as an adult, can grow to 60–100 cm (2–3 ft) long and be as thick as a spaghetti noodle.

Infection does not create immunity, so people can repeatedly experience Dracunculiasis throughout their lives.

In drier areas just south of the Sahara desert, cases of the disease often emerge during the rainy season, which for many agricultural communities is also the planting or harvesting season. Elsewhere, the emerging worms are more prevalent during the dry season, when ponds and lakes are smaller and copepods are thus more concentrated in them. Guinea worm disease outbreaks can cause serious disruption to local food supplies and school attendance.

The infection can be acquired by eating a fish paratenic host, but this is rare. No reservoir hosts are known; that is, each generation of worms must pass through a human – or possibly a dog.

Prevention

Guinea worm disease can be transmitted only by drinking contaminated water, and can be completely prevented through two relatively simple measures:

1. Prevent people from drinking contaminated water containing the Cyclops copepod (water flea), which can be seen in clear water as swimming white specks.

    o Drink water drawn only from sources free from contamination.

    o Filter all drinking water, using a fine-mesh cloth filter like nylon, to remove the guinea worm-containing crustaceans. Regular cotton cloth folded over a few times is an effective filter.

    o Filter the water through ceramic or sand filters.

    o Boil the water.

    o Develop new sources of drinking water without the parasites, or repair dysfunctional water sources.

    o Treat water sources with larvicides to kill the water fleas.

2. Prevent people with emerging Guinea worms from entering water sources used for drinking.

    o Community-level case detection and containment is key. For this, staff must go door to door looking for cases, and the population must be willing to help and not hide their cases.

- o Immerse emerging worms in buckets of water to reduce the number of larvae in those worms, and then discard that water on dry ground.

- o Discourage all members of the community from setting foot in the drinking water source.

- o Guard local water sources to prevent people with emerging worms from entering.

Treatment

There is no vaccine or medicine to treat or prevent Guinea worm disease. Once a Guinea worm begins emerging, the first step is to do a controlled submersion of the affected area in a bucket of water. This causes the worm to discharge many of its larva, making it less infectious. The water is then discarded on the ground far away from any water source. Submersion results in subjective relief of the burning sensation and makes subsequent extraction of the worm easier. To extract the worm, a person must wrap the live worm around a piece of gauze or a stick. The process can be long, taking anywhere from hours to a week. Gently massaging the area around the blister can help loosen the worm. This is nearly the same treatment that is noted in the famous ancient Egyptian medical text, the Ebers papyrus from 1550 BC. Some people have said that extracting a Guinea worm feels like the afflicted area is on fire. However, if the infection is identified before an ulcer forms, the worm can also be surgically removed by a trained doctor in a medical facility.

Although Guinea worm disease is usually not fatal, the wound where the worm emerges could develop a secondary bacterial infection such as tetanus, which may be life-threatening—a concern in endemic areas where there is typically limited or no access to health care. Analgesics can be used to help reduce swelling and pain and antibiotic ointments can help prevent secondary infections at the wound site. At least in the Northern region of Ghana, the Guinea worm team found that antibiotic ointment on the wound site caused the wound to heal too well and too quickly making it more difficult to extract the worm and more likely that pulling would break the worm. The local team preferred to use something called "Tamale oil" (after the regional capital) which lubricated the worm and aided its extraction.

It is of great importance not to break the worm when pulling it out. Broken worms have a tendency to putrefy or petrify. Putrefaction leads to the skin sloughing off around the worm. Petrification is a problem if the worm is in a joint or wrapped around a vein or other important area.

Use of metronidazole or thiabendazole may make extraction easier, but also may lead to migration to other parts of the body.

Epidemiology

In 1986, there were an estimated 3.5 million cases of Guinea worm in 20 endemic nations in Asia and Africa. Ghana alone reported 180,000 cases in 1989. The number of cases has since been reduced by more than 99.999% to 22 in 2015 — in the four remaining endemic nations of Africa: South Sudan, Chad, Mali and Ethiopia. This is the lowest number of cases since the eradication campaign began. As of 2010, however, the WHO predicted it will be "a few years yet" before eradication is achieved, on the basis that it took 6–12 years for the countries that have so far eliminated Guinea worm transmission to do so after reporting a similar number of cases to that reported in southern Sudan (now South Sudan) in 2009.

The World Health Organization is the international body that certifies whether a disease has been eliminated from a country or eradicated from the world. Former U.S. President Jimmy Carter's not-for-profit organization, the Carter Center, also reports the status of the Guinea worm eradication program by country.

Certified Free

Endemic countries must report to the International Commission for the Certification of Dracunculiasis Eradication and document the absence of indigenous cases of Guinea worm disease for at least three consecutive years to be certified as Guinea worm-free by the World Health Organization.

The results of this certification scheme have been remarkable: by 2007, Benin, Burkina Faso, Chad, Côte d'Ivoire, Kenya, Mauritania, Togo, and Uganda had stopped transmission, and Cameroon, Central African Republic, India, Pakistan, Senegal, Yemen were WHO certified. Nigeria was certified as having ended transmission in 2013, followed by Ghana in 2015.

Endemic

With the current eradication campaign the areas that dracunculiasis are found are shrinking. In the early 1980s, the disease was endemic in Pakistan, Yemen and 17 countries in Africa with a total of 3.5 million cases per year. In 1985, 3.5 million cases were still reported annually, but by 2008, the number had dropped to 5,000. This number further dropped to 1058 in 2011. At the end of 2013, South Sudan, Mali, Ethiopia and Chad still had endemic transmission. For many years the major focus was South Sudan (independent after 2011, formerly the southern region of Sudan), which reported 76% of all cases in 2013. Now all 4 countries with endemic cases look close to eliminating the disease.

Date	South Sudan	Mali	Ethiopia	Chad	Total
2011	1,028	12	8	10	**1058**
2012	521	7	4	10	**542**
2013	113	11	7	14	**148** (including 3 exported to Sudan)
2014	70	40	3	13	**126**
2015	5	5	3	9	**22**
2016 - up to 30 June	2	0	1	4	7 (provisional)

Eradication Program

In 1984, the WHO asked the United States Centers for Disease Control and Prevention (CDC) to spearhead the effort to eradicate dracunculiasis, an effort that was further supported by the Carter Center, former U.S. President Jimmy Carter's not-for-profit organization. In 1986, Carter and the Carter Center began leading the global campaign, in conjunction with CDC, UNICEF, and WHO. At that time the disease was endemic in Pakistan, Yemen and 17 countries in Africa, which reported a total of 3.5 million cases per year.

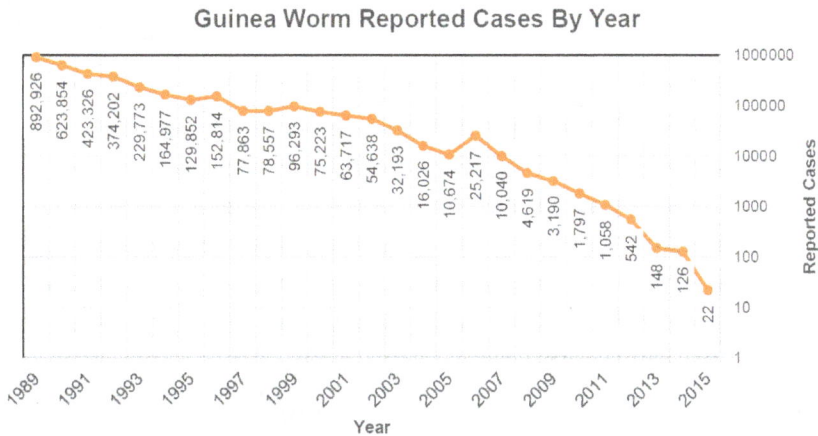

Guinea Worm Reported Cases By Year

892,926 623,854 423,326 374,202 229,773 164,977 129,852 152,814 77,863 78,557 96,293 75,223 63,717 54,638 32,193 16,026 10,674 25,217 10,040 4,619 3,190 1,797 1,058 542 148 126 22

Logarithmic scale of reported human cases of guinea worm by year, 1989–2015 (2015 data is provisional). Data from Guinea Worm Eradication Program.

Since humans are the principal host for Guinea worm, and there is no evidence that *D. medinensis* has ever been reintroduced to humans in any formerly endemic country as the result of non-human infections, the disease can be controlled by identifying all cases and modifying human behavior to prevent it from recurring. Once all human cases are eliminated, the disease cycle will be broken, resulting in its eradication.

The eradication of Guinea worm disease has faced several challenges:

- Inadequate security in some endemic countries

- Lack of political will from the leaders of some of the countries in which the disease is endemic

- The need for change in behavior in the absence of a magic bullet treatment like a vaccine or medication

- Inadequate funding at certain times

Carter made a personal visit to a Guinea-worm endemic village in 1988. He said, "Encountering those victims first-hand, particularly the teenagers and small children, propelled me and Rosalynn [his wife] to step up the Carter Center's efforts to eradicate Guinea worm disease."

In 1991, the World Health Assembly (WHA) agreed that Guinea worm disease should be eradicated. At this time there were 400,000 cases reported each year. The Carter Center has continued to lead the eradication efforts, primarily through its Guinea Worm Eradication Program.

In the 1980s, Carter persuaded President Zia al-Haq of Pakistan to accept the proposal of the eradication program, and by 1993, Pakistan was free of the disease. Key to the effort was, according to Carter, the work of "village volunteers" who educated people about the need to filter drinking water. Other countries followed, and by 2004, the worm was eradicated in Asia.

In December 2008, The Carter Center announced new financial support totaling $55 million from the Bill & Melinda Gates Foundation and the United Kingdom Department for International Development. The funds will help address the higher cost of identifying and reporting the last cases of Guinea worm disease. Since the worm has a one-year incubation period, there is a very high cost of maintaining a broad and sensitive monitoring system and providing a rapid response when necessary.

One of the most significant challenges facing Guinea worm eradication has been the civil war in southern Sudan, which was largely inaccessible to health workers due to violence. To address some of the humanitarian needs in southern Sudan, in 1995, the longest ceasefire in the history of the war, and the longest humanitarian cease-fire in history, was achieved through negotiations by Jimmy Carter. Commonly called the "Guinea worm cease-fire," both warring parties agreed to halt hostilities for nearly six months to allow public health officials to begin Guinea worm eradication programming, among other interventions.

Public health officials cite the formal end of the war in 2005 as a turning point in Guinea worm eradication because it has allowed health care workers greater access to southern Sudan's endemic areas. In 2006, there was an increase to 15,539 cases, from 5,569 cases in 2005, as a result of better reporting from areas that were no longer war-torn. The Southern Sudan Guinea Worm Eradication Program (SSGWEP) has deployed over 28,000 village volunteers, supervisors and other health staff to work on the program full-time. The SSGWEP was able to slash the number of cases reported in 2006 by 63% to 5,815 cases in 2007. Since 2011, at the time that South Sudan became independent, its northern neighbor Sudan had reported no endemic cases of dracunculiasis .

Sporadic insecurity or widespread civil conflict could at any time ignite, thwarting eradication efforts. The remaining endemic communities in South Sudan are remote, poor and devoid of infrastructure, presenting significant hurdles for effective delivery of interventions against disease. Moreover, residents in these communities are nomadic, moving seasonally with cattle in pursuit of water and pasture, making it very difficult to know where and when transmission occurred. The peak transmission season coincides with the rainy season, hampering travel by public health workers.

One remaining area in West Africa outside of Ghana remains challenging to ending Guinea worm: northern Mali, where Tuareg rebels have made some affected areas unsafe for health workers. Four of Mali's regions—(Kayes, Koulikoro, Ségou, and Sikasso)—have eliminated dracunculiasis, while the disease is still endemic in the country's other four regions (Gao, Kidal, Mopti, and Timbuktu). Late detection of two outbreaks, due to inadequate surveillance resulted in a meager 36% containment rate in Mali in 2007. The years 2008 and 2009 were more successful, however, with containment rates of 85% and 73% respectively. The civil war prevented accurate information from being gathered in northern Mali in 2012.

From June 2006 to March 2008, there had been no cases reported in Ethiopia.

Before 2010, Chad had not reported any indigenous cases of guinea worm in over 10 years.

In Ghana, after a decade of frustration and stagnation, in 2006 a decisive turnaround was achieved. Multiple changes can be attributed to the improved containment and lower incidence of dracunculiasis: better supervision and accountability, active oversight of infected people daily by paid staff, and an intensified public awareness campaign. After Jimmy Carter's visit to Ghana in August 2006, the government of Ghana declared Guinea worm disease to be a public health emergency. The overall rate of contained cases has increased in Ghana from 60% in 2005, to 75% in 2006, 84% in 2007, 85% in 2008, 93% in 2009, and 100% in 2010.

On 30 January 2012 the WHO meeting at the Royal College of Physicians in London launched the most ambitious and largest coalition health project ever, known as *London Declaration on Neglected Tropical Diseases* which aims to end/control dracunculiasis by 2020, among other neglected tropical disease. This project is supported of all major pharmaceutical companies, the Bill & Melinda Gates Foundation, the governments of the United States, United Kingdom DFID and United Arab Emirates and the World Bank.

In 2015 22 cases of dracunculiasis were reported: 9 in Chad, 3 in Ethiopia, 5 in Mali and 5 in South Sudan. This is an 83% reduction from 2014. The proportion of people contained (i.e. treated and isolated from drinking water sources early enough to remove the risk they can contaminate the water source) is 36%, compared to 73% in 2014. That means 14 cases have not been contained in 2015, compared to 34 cases in 2014. 9 of these 14 cases not contained were in Chad.

A significant change from 2014 is the increased effort being used to identify and treat infected dogs—mainly in Chad where the vast majority of cases of dogs hosting the worm have been found, but also significantly in Ethiopia. In 2015, 483 infected dogs were identified and treated in Chad — more than 20 times the number reported in humans worldwide. This is more than 4 times larger than the number treated in 2014 (114 dogs). A major factor in this increase is probably the financial reward started in January for reporting an infected dog. 68% of dogs treated were also contained, compared to 40% in 2014. Dogs are believed to be the major source of the parasite infecting humans in Chad. 15 dogs outside Chad have also been identified and treated, as well 5 cats and 1 baboon. The August Carter Center report predicts that Chad may be the last country that eliminates dracunculiasis, and reports on further ongoing research into the relationship between the parasite and dogs there, and some different treatments for dogs. It also predicts that the large increase in monitoring, treating and containing dogs this year will not affect the number of human cases for many months due to the 1 year incubation period of the disease.

In August 2015, when discussing his diagnosis of melanoma metastasized to his brain, Jimmy Carter stated that he hopes the last Guinea worm dies before he does.

In 2016 4 cases have been reported up to 31 May - 3 in Chad, 1 in Ethiopia - compared to 5 cases last year in the same period - 4 in Chad, 1 in Ethiopia. Mali and South Sudan have not reported any new cases in 6 months - although 2 possible cases in South Sudan are still being investigated. The efforts against infected dogs continues to increase in Chad, with 498 dogs being identified and treated up to 31 May, compared to 196 cases in the same period the previous year. The level of containment of infected dogs before they become a risk of spreading the parasite has improved to 80% compared to 67% last year - based on containment figures up to 30 April 2016.

Society and Culture

The pain caused by the worm's emergence—which typically occurs during planting and harvesting seasons—prevents many people from working or attending school for as long as three months. In heavily burdened agricultural villages fewer people are able to tend their fields or livestock, resulting in food shortages and lower earnings. A study in southeastern Nigeria, for example, found that rice farmers in a small area lost US$20 million in just one year due to outbreaks of Guinea worm disease.

History

Dracunculiasis has been a recognized disease for thousands of years:

- Guinea worm has been found in calcified Egyptian mummies.

- An Old Testament description of "fiery serpents" may have been referring to Guinea worm: "And the Lord sent fiery serpents among the people, and they bit the people; and much people of Israel died." (Numbers 21:4–9).

- The 2nd century BC, Greek writer Agatharchides, described this affliction as being endemic amongst certain nomads in what is now Sudan and along the Red Sea.

- The unusually high incidence of dracunculiasis in the city of Medina led to it being included in part of the disease's scientific name "medinensis." A similar high incidence along the Guinea coast of West Africa gave the disease its more commonly used name. Guinea worm is no longer endemic in either location.

The Russian scientist Alexei Pavlovich Fedchenko (1844–1873) during the 1860s while living in Samarkand was provided with a number of specimens of the worm by a local doctor which he kept in water. While examining the worms Fedchenko noted the presence of water fleas with embryos of the guinea worm within them.

In modern times, the first to describe dracunculiasis and its pathogenesis was the Bulgarian physician Hristo Stambolski, during his exile in Yemen (1877–1878). His theory was that the cause was infected water which people were drinking.

Etymology

Dracunculiasis once plagued a wide band of tropical countries in Africa and Asia. Its Latin name, *Dracunculus medinensis* ("little dragon from Medina"), derives from its one-time high incidence in the city of Medina, and its common name, Guinea worm, is due to a similar past high incidence along the Guinea coast of West Africa; both of these locations are now free of Guinea worm. In the 18th century, Swedish naturalist Carl Linnaeus identified *D. medinensis* in merchants who traded along the Gulf of Guinea (West African Coast), hence the name Guinea worm.

Other Animals

A very similar or possibly the same worm has been found in dogs. It is unclear if dog and human infections are related. It is possible that dogs may be able to spread the disease to people, that a third organism may be able to spread it to both dogs and people, or that this may be a different type of *Dracunculus*.

Enterobiasis

A pinworm infection or enterobiasis is a human parasitic disease and one of the most common parasitic worm infections (also called helminthiasis) in the developed world. It is caused by infestation of the parasitic roundworm *Enterobius vermicularis*, commonly called the human *pinworm*. Infection usually occurs through the ingestion of pinworm eggs, either through contaminated hands, food, or less commonly, water. The chief symptom is itching in the anal area. The incubation time from ingestion of eggs to the first appearance of new eggs around the anus is 4 to 6 weeks.

Pinworms are usually considered a nuisance rather than a serious disease. For this reason, enterobiasis is not classified as a neglected tropical disease unlike many other parasitic worm infections.

Treatment is straightforward in uncomplicated cases; however, elimination of the parasite from a family group or institution often poses significant problems—either due to an incomplete cure or reinfection.

Signs and Symptoms

One third of individuals with pinworm infection are totally asymptomatic. The main symptoms are pruritus ani and perineal pruritus, i.e., itching in and around the anus and around the perineum. The itching occurs mainly during the night, and is caused by the female pinworms migrating to lay eggs around the anus. Both the migrating females and the clumps of eggs are irritating, but the mechanisms causing the intense pruritus have not been explained. The intensity of the itching varies, and it can be described as tickling, crawling sensations, or even acute pain. The itching leads to continuously scratching the area around the anus, which can further result in tearing of the skin and complications such as secondary bacterial infections, including bacterial dermatitis (i.e., skin inflammation) and folliculitis (i.e., hair follicle inflammation). General symptoms are insomnia (i.e., persistent difficulties to sleep) and restlessness. A considerable proportion of children suffer from loss of appetite, weight loss, irritability, emotional instability, and enuresis (i.e., inability to control urination).

Two female pinworms next to a ruler. The markings are one millimeter apart.

Pinworms cannot damage the skin, and they do not normally migrate through tissues. However, in women they may move onto the vulva and into the vagina, from there moving to the external orifice of the uterus, and onwards to the uterine cavity, fallopian tubes, ovaries, and peritoneal cavity. This can cause vulvovaginitis, i.e. an inflammation of the vulva and vagina. This causes vaginal discharge and pruritus vulvae, i.e., itchiness of the vulva. The pinworms can also enter the urethra, and presumably, they carry intestinal bacteria with them. According to Gutierrez (2000), a statistically significant correlation between pinworm infection and urinary tract infections has been shown; however, Burkhart & Burkhart (2005) maintain that the incidence of pinworms as a cause of urinary tract infections remains unknown. Incidentally, one report indicated that 36% of young girls with a urinary tract infection also had pinworms. Dysuria (i.e., painful urination) has been associated with pinworm infection.

The relationship between pinworm infestation and appendicitis has been researched, but there is a lack of clear consensus in the matter: while Gutierres (2005) maintains that there exists a consensus that pinworms do not produce the inflammatory reaction, Cook (1994) states that it is controversial whether pinworms are causatively related to acute appendicitis, and Burkhart & Burkhart (2004) state that pinworm infection causes symptoms of appendicitis to surface.

Cause

The cause of a pinworm infection is the worm *Enterobius vermicularis*. The entire lifecycle — from egg to adult — takes place in the human gastrointestinal tract of a single human host. Cook *et al.* (2009) and Burkhart & Burkhart (2005) disagree over the length of this process, with Cook *et al.* stating two to four weeks, while Burkhart & Burkhart states that it takes from four to eight weeks.

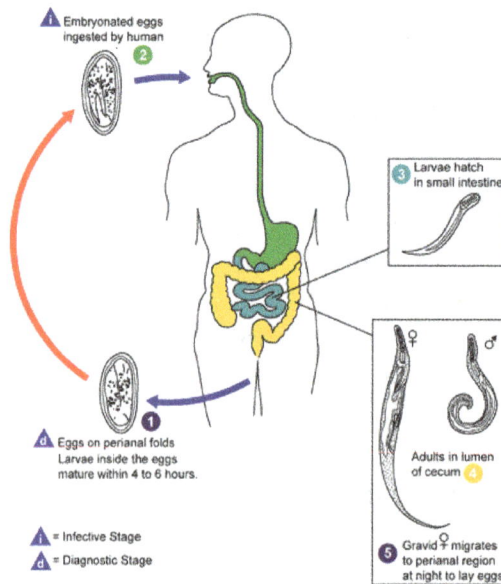

Pinworm life cycle.

The lifecycle begins with eggs being ingested. The eggs hatch in the duodenum (i.e., first part of the small intestine). The emerging pinworm larvae grow rapidly to a size of 140 to 150 micrometers in size, and migrate through the small intestine towards the colon. During this migration they moult twice and become adults. Females survive for 5 to 13 weeks, and males about 7 weeks. The male and female pinworms mate in the ileum (i.e., last part of the small intestine), whereafter the male pinworms usually die, and are passed out with stool. The gravid female pinworms settle in the ileum, caecum (i.e., beginning of the large intestine), appendix and ascending colon, where they attach themselves to the mucosa and ingest colonic contents. Almost the entire body of a gravid female becomes filled with eggs. The estimations of the number of eggs in a gravid female pinworm ranges from about 11,000 to 16,000. The egg-laying process begins approximately five weeks after initial ingestion of pinworm eggs by the human host. The gravid female pinworms migrate through the colon towards the rectum at a rate of 12 to 14 centimeters per hour. They emerge from the anus, and while moving on the skin near the anus, the female pinworms deposit eggs either through (1) contracting and expelling the eggs, (2) dying and then disintegrating, or (3) bodily rupture due to the host scratching the worm. After depositing the eggs, the female becomes opaque and dies. The reason the female emerges from the anus is to obtain the oxygen necessary for the maturation of the eggs.

Pinworm infection spreads through human-to-human transmission, by ingesting (i.e., swallowing) infectious pinworm eggs. The eggs are hardy and can remain viable (i.e., infectious) in a moist environment for up to three weeks. They do not tolerate heat well, but can survive in low temperatures: two-thirds of the eggs are still viable after 18 hours at −8 degrees Celsius (18 °F).

After the eggs have been initially deposited near the anus, they are readily transmitted to other surfaces through contamination. The surface of the eggs is sticky when laid, and the eggs are readily transmitted from their initial deposit near the anus to fingernails, hands, night-clothing and bed linen. From here, eggs are further transmitted to food, water, furniture, toys, bathroom fixtures and other objects. Household pets often carry the eggs in their fur, while not actually being infected. Dust containing eggs can become airborne and widely dispersed when dislodged from surfaces, for instance when shaking out bed clothes and linen. Consequently, the eggs can enter the mouth and nose through inhalation, and be swallowed later. Although pinworms do not strictly multiply inside the body of their human host, some of the pinworm larvae may hatch on the anal mucosa, and migrate up the bowel and back into the gastrointestinal tract of the original host. This process is called *retroinfection*. According to Burkhart (2005), when this retroinfection occurs, it leads to a heavy parasitic load and ensures that the pinworm infestation continues. This statement is contradictory to a statement by Caldwell, who contends that retroinfection is rare and not clinically significant. Despite the limited, 13 week lifespan of individual pinworms, autoinfection (i.e., infection from the original host to itself), either through the anus-to-mouth route or through retroinfection, causes the pinworms to inhabit the same host indefinitely.

Diagnosis

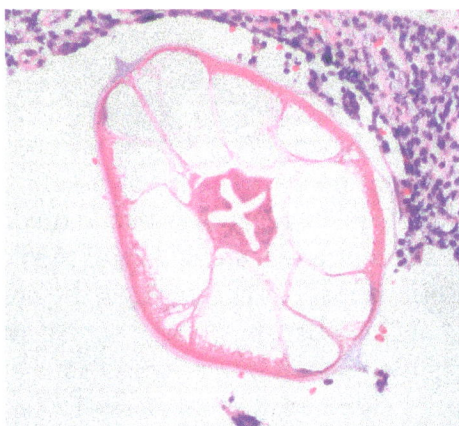

High magnification micrograph of a pinworm in cross section in the appendix. H&E stain.

Diagnosis depends on finding the eggs or the adult pinworms. Individual eggs are invisible to the naked eye, but they can be seen using a low-power microscope. On the other hand, the light-yellowish thread-like adult pinworms are clearly visually detectable, usually during the night when they move near the anus, or on toilet paper. Transparent adhesive tape (e.g. Scotch Tape) applied on the anal area will pick up deposited eggs, and diagnosis can be made by examining the tape with a microscope. This test is most successful if done every morning for several days, because the females do not lay eggs every day, and the number of eggs vary.

Pinworms do not lay eggs in the feces, but sometimes eggs are deposited in the intestine. As such, routine examination of fecal material gives a positive diagnosis in only 5 to 15% of infected subjects, and is therefore of little practical diagnostic use. In a heavy infection, female pinworms may adhere to stools that pass out through the anus, and they may thus be detected on the surface on the stool. Adult pinworms are occasionally seen during colonoscopy. On a microscopic level, pinworms have an identifying feature of alae (i.e., protruding ridges) running the length of the worm.

Prevention

Pinworm infection cannot be totally prevented under most circumstances. This is due to the prevalence of the parasite and the ease of transmission through soiled night clothes, airborne eggs, contaminated furniture, toys and other objects. Infection may occur in the highest strata of society, where hygiene and nutritional status are typically high. The stigma associated with pinworm infection is hence considered a possible over-emphasis. Counselling is sometimes needed for upset parents that have discovered their children are infected, as they may not realize how prevalent the infection is.

Preventative action revolves around personal hygiene and the cleanliness of the living quarters. The *rate* of reinfection can be reduced through hygienic measures, and this is recommended especially in recurring cases. The main measures are keeping fingernails short, and washing and scrubbing hands and fingers carefully, especially after defecation and before meals. Under ideal conditions, bed covers, sleeping garments, and hand towels should be changed daily. Simple laundering of clothes and linen disinfects them. Children should wear gloves while asleep, and the bedroom floor should be kept clean. Food should be covered to limit contamination with dust-borne parasite eggs. Household detergents have little effect on the viability of pinworm eggs, and cleaning the bathroom with a damp cloth moistened with an antibacterial agent or bleach will merely spread the still-viable eggs. Similarly, shaking clothes and bed linen will detach and spread the eggs.

Treatment

Medication is the primary treatment for pinworm infection. The existing pharmaceutical drugs against pinworms are so effective that many medical scientists regard hygienic measures as impractical. However, reinfection is frequent regardless of the medication used. Total elimination of the parasite in a household may require repeated doses of medication for up to a year or more. Because the drugs kill the adult pinworms, but not the eggs, the first retreatment is recommended in two weeks. Also, if one household member spreads the eggs to another, it will be a matter of two or three weeks before those eggs become adult worms and thus amenable to treatment. Asymptomatic infections, often in small children, can serve as reservoirs of infection, and therefore the entire household should be treated regardless of whether or not symptoms are present.

The benzimidazole compounds albendazole (brand names e.g., *Albenza, Eskazole, Zentel* and *Andazol*) and mebendazole (brand names e.g., *Ovex, Vermox, Antiox* and *Pripsen*) are the most effective. They work by inhibiting the microtubule function in the pinworm adults, causing glycogen depletion, thereby effectively starving the parasite. A single 100 milligram dose of mebendazole with one repetition after a week, is considered the safest, and is usually effective with cure rate of 96%. Mebendazole has no serious side effects, although abdominal pain and diarrhea have been reported. Pyrantel pamoate (also called pyrantel embonate, brand names e.g., *Reese's Pinworm Medicine, Pin-X, Combantrin, Anthel, Helmintox*, and *Helmex*) kills adult pinworms through neuromuscular blockade, and is considered as effective as the benzimidazole compounds and is used as a second-line medication. Other medications are piperazine, which causes flaccid paralysis in the adult pinworms, and pyrvinium pamoate (also called pyrvinium embonate), which works by inhibiting oxygen uptake of the adult pinworms. Pinworms located in the genitourinary system (in this case, female genital area) may require other drug treatments.

Garlic has been used as a treatment throughout history in the ancient cultures of China, India, Egypt, and Greece. Hippocrates (459–370 BC) mentioned garlic as a remedy against intestinal parasites. German botanist Lonicerus (1564) recommended garlic against parasitic worms. The action of garlic is manifold. Because of allicin and other sulfur compounds, garlic has antibiotic, antibacterial and antimycotic action, which has been testified by in vitro studies. It can be concluded that administration of garlic should not be avoided; on the contrary, its intake should be as much as possible. Crushed garlic cloves and tea tree oil mixture can be applied (over isolation layer like vaseline) to the anus to stop itching. Applying raw garlic on skin may cause severe chemical burn.

Epidemiology

Pinworm infection occurs worldwide, and is the most common helminth (i.e., parasitic worm) infection in the United States and Western Europe. In the United States, a study by the Center of Disease Control reported an overall incidence rate of 11.4% among people of all ages. Pinworms are particularly common in children, with prevalence rates in this age group having been reported as high as 61% in India, 50% in England, 39% in Thailand, 37% in Sweden, and 29% in Denmark. Finger sucking has been shown to increase both incidence and relapse rates, and nail biting has been similarly associated. Because it spreads from host to host through contamination, enterobiasis is common among people living in close contact, and tends to occur in all people within a household. The prevalence of pinworms is not associated with gender, nor with any particular social class, race, or culture. Pinworms are an exception to the tenet that intestinal parasites are uncommon in affluent communities. The earliest known instance of pinworms is evidenced by pinworm eggs found in coprolite, carbon dated to 7837 BC at western Utah.

References

- Farrar, Jeremy; Hotez, Peter; Junghanss, Thomas; Kang, Gagandeep; Lalloo, David; White, Nicholas J. (2013-10-26). Manson's Tropical Diseases. Elsevier Health Sciences. pp. 664–671. ISBN 9780702053061.

- Day, David W.; Basil C. Morson; Jeremy R. Jass; Geraint Williams; Ashley B. Price (2003). Morson and Dawson's Gastrointestinal Pathology. John Wiley & Sons, Inc. ISBN 978-0-632-04204-3.

- James, William D.; Berger, Timothy G.; et al. (2006). Andrews' Diseases of the Skin: clinical Dermatology. Saunders Elsevier. ISBN 0-7216-2921-0.

- "Trematodes (Schistosomes and Liver, Intestinal, and Lung Flukes)". Mandell, Douglas, and Bennett's Principles and Practice of Infectious Diseases. pp. 3216–3226.e3. ISBN 0443068399.

- WHO (2006). Guidelines for the Safe Use of Wastewater, Excreta and Greywater, Volume 4 Excreta and Greywater Use in Agriculture. (third ed.). Geneva: World Health Organization. ISBN 9241546859.

- Luke F. Pennington and Michael H. Hsieh (2014) Immune Response to Parasitic Infections, Bentham e books, Vol 2, pp. 93-124, ISBN 978-1-60805-148-9

- Junghanss, Jeremy Farrar, Peter J. Hotez, Thomas (2013). Manson's tropical diseases. (23rd ed.). Oxford: Elsevier/Saunders. p. e62. ISBN 9780702053061.

- G. D. Schmidt; L S. Roberts (2009). Larry S. Roberts; John Janovy, Jr., eds. Foundations of Parasitology (8th ed.). McGraw-Hill. pp. 480–484. ISBN 978-0-07-128458-5.

- Gutiérrez, Yezid (2000). Diagnostic pathology of parasitic infections with clinical correlations (PDF) (Second ed.). Oxford University Press. pp. 354–366. ISBN 0-19-512143-0. Retrieved 21 August 2009.

- Garcia, Lynne Shore (2009). Practical guide to diagnostic parasitology. American Society for Microbiology. pp. 246–247. ISBN 1-55581-154-X. Retrieved 2009-12-05.

- "Cryptosporidium: Illness & Symptoms". United States Centers for Disease Control and Prevention. 20 February 2015. Retrieved 11 January 2016.

- ""Number of Reported Cases of Guinea Worm Disease by Year: 1989–2015*"" (PDF). Guinea Worm Eradication Program. 2016-01-06. Retrieved 2016-01-10.

- "Cryptosporidium: Sources of Infection & Risk Factors". United States Centers for Disease Control and Prevention. 1 April 2015. Retrieved 16 January 2016.

- Uniting to Combat Neglected Tropical Diseases (30 January 2012). "London Declaration on Neglected Tropical Diseases" (PDF). Uniting to Combat NTDs. Retrieved 2013-05-06.

- WHO (3 February 2012). "WHO roadmap inspires unprecedented support to defeat neglected tropical diseases". World Health Organization. Retrieved 2013-05-06.

- Haque, Rashidul; Huston, Christopher D.; Hughes, Molly; Houpt, Eric; Petri, William A. (2003-04-17). "Amebiasis". NEJM. 348 (16): 1565–1573. doi:10.1056/NEJMra022710. Retrieved 2012-04-12.

8

Aquatic Toxicology: An Overview

The study of the harm done to the aquatic organisms by humans is known as aquatic toxicology. The fields concerned with aquatic toxicology are freshwater, marine water and sediment environments. Aquatic toxicology is an emerging field of study, the following chapter will not only provide an overview but will also delve into the topics related to it.

Aquatic Toxicology

Aquatic toxicology is the study of the effects of manufactured chemicals and other anthropogenic and natural materials and activities on aquatic organisms at various levels of organization, from subcellular through individual organisms to communities and ecosystems. Aquatic toxicology is a multidisciplinary field which integrates toxicology, aquatic ecology and aquatic chemistry.

This field of study includes freshwater, marine water and sediment environments. Common tests include standardized acute and chronic toxicity tests lasting 24–96 hours (acute test) to 7 days or more (chronic tests). These tests measure endpoints such as survival, growth, reproduction, that are measured at each concentration in a gradient, along with a control test. Typically using selected organisms with ecologically relevant sensitivity to toxicants and a well-established literature background. These organisms can be easily acquired or cultured in lab and are easy to handle.

History

While basic research in toxicology began in multiple countries in the 1800s, it was not until around the 1930s that the use of acute toxicity testing, especially on fish, was established. Over the next two decades, the effects of chemicals and wastes on non-human species became more of a public issue and the era of the *pickle-jar bioassays* began as efforts increased to standardize toxicity testing techniques. In the United States of America, the passage of the Federal Water Pollution Control Act of 1947 marked the first comprehensive legislation for the control of water pollution and was followed by the Federal Water Pollution Control Act in 1956. In 1962, public and governmental interests were renewed, in large part due to the publication of Rachel Carson's *Silent Spring*, and three years later the Water Quality Act was passed which directed states to develop water quality standards. Public awareness, as well as scientific and governmental concern, continued to grow throughout the 1970s and by the end of the decade research had expanded to include hazard evaluation and risk analysis. In the subsequent decades, aquatic toxicology has continued to expand and internationalize so that there is now a strong application of toxicity testing for environmental protection.

Aquatic Toxicity Tests

Aquatic toxicology tests (assays): toxicity tests are used to provide qualitative and quantitative data on adverse (deleterious) effects on aquatic organisms from a toxicant. Toxicity tests can be used to assess the potential for damage to an aquatic environment and provide a database that can be used to assess the risk associated within a situation for a specific toxicant. Aquatic toxicology tests can be performed in the field or in the laboratory. Field experiments generally refer to multiple species exposure and laboratory experiments generally refer to single species exposure. A dose response relationship is most commonly used with a sigmoidal curve to quantify the toxic effects at a selected end-point or criteria for effect (i.e. death or other adverse effect to the organism). Concentration is on the x-axis and percent inhibition or response is on the y-axis.

The criteria for effects, or endpoints tested for, can include lethal and sublethal effects.

There are different types of toxicity tests that can be performed on various test species. Different species differ in their susceptibility to chemicals, most likely due to differences in accessibility, metabolic rate, excretion rate, genetic factors, dieteary factors, age, sex, health and stress level of the organism. Common standard test species are the fathead minnow (Pimephales promelas), daphnids (*Daphnia magna, D. pulex, D. pulicaria, Ceriodaphnia dubia*), midge (Chironomus tentans, C. ruparius), rainbow trout (Oncorhynchus mykiss), sheepshead minnow (Cyprinodon variegatu), mysids (Mysidopsis), oyster (Crassotreas), scud (Hyalalla Azteca), grass shrimp (Palaemonetes pugio), mussels (Mytilus). As defined by ASTM, these species are routinely selected on the basis of availability, commercial, recreational, and ecological importance, past successful use, and regulatory use.

A variety of acceptable standardized test methods have been published. Some of the more widely accepted agencies to publish methods are: the American Public Health Association, U.S. Environmental Protection Agency, American Society for Testing and Materials, International Organization for Standardization, Environment Canada, and Organization for Economic Cooperation and Development. Standardized tests offer the ability to compare results between laboratories.

There are many kinds of toxicity tests widely accepted in the scientific literature and regulatory agencies. The type of test used depends on many factors: Specific regulatory agency conducting the test, resources available, physical and chemical characteristics of the environment, type of toxicant, test species available, laboratory vs. field testing, end-point selection, and time and resources available to conduct the assays are some of the most common influencing factors on test design.

Exposure Systems

Exposure systems are four general techniques the controls and test organisms are exposed to the dealing with treated and diluted water or the test solutions.

Static- a static test exposes the organism in still water. The toxicant is added to the water in order to obtain the correct concentrations to be tested. The control and test organisms are placed in the test solutions and the water is not changed for the entirety of the test.

Recirculation- a recirculation test exposes the organism to the toxicant in a similar manner as the static test, except that the test solutions are pumped through an apparatus (i.e. filter) to maintain water quality, but not reduce the concentration of the toxicant in the water. The water is circulated through the test chamber continuously, similar to an aerated fish tank. This type of test is expensive and it is unclear whether or not the filter or aerator has an effect on the toxicant.

Renewal- a renewal test also exposes the organism to the toxicant in a similar manner as the static test because it is in still water. However, in a renewal test the test solution is renewed periodically (constant intervals) by transferring the organism to a fresh test chamber with the same concentration of toxicant.

Flow-through- a flow through test exposes the organism to the toxicant with a flow into the test chambers and then out of the test chambers. The once-through flow can either be intermittent or continuous. A stock solution of the correct concentrations of contaminant must be previously prepared. Metering pumps or diluters will control the flow and the volume of the test solution, and the proper proportions of water and contaminant will be mixed.

Types of Tests

Acute tests are short-term exposure tests (hours or days) and generally use lethality as an endpoint. In acute exposures, organisms come into contact with higher doses of the toxicant in a single event or in multiple events over a short period of time and usually produce immediate effects, depending on absorption time of the toxicant. These tests are generally conducted on organisms during a specific time period of the organism's life cycle, and are considered partial life cycle tests. Acute tests are not valid if mortality in the control sample is greater than 10%. Results are reported in EC50, or concentration that will affect fifty percent of the sample size.

Chronic tests are long-term tests (weeks, months years), relative to the test organism's life span (>10% of life span), and generally use sub-lethal endpoints. In chronic exposures, organisms come into contact with low, continuous doses of a toxicant. Chronic exposures may induce effects to acute exposure, but can also result in effects that develop slowly. Chronic tests are generally considered full life cycle tests and cover an entire generation time or reproductive life cycle ("egg to egg"). Chronic tests are not considered valid if mortality in the control sample is greater than 20%. These results are generally reported in NOECs (No observed effects level) and LOECs (Lowest observed effects level).

Early life stage tests are considered as subchronic exposures that are less than a complete reproductive life cycle and include exposure during early, sensitive life stages of an organism. These exposures are also called critical life stage, embryo-larval, or egg-fry tests. Early life stage tests are not considered valid if mortality in the control sample is greater than 30%.

Short-term sublethal tests are used to evaluate the toxicity of effluents to aquatic organisms. These methods are developed by the EPA, and only focus on the most sensitive life stages. Endpoints for these test include changes in growth, reproduction and survival. NOECs, LOECs and EC50s are reported in these tests.

Bioaccumulation tests are toxicity tests that can be used for hydrophobic chemicals that may accumulated in the fatty tissue of aquatic organisms. Toxicants with low solubilities in water generally can be stored in the fatty tissue due to the high lipid content in this tissue. The storage of these toxicants within the organism may lead to cumulative toxicity. Bioaccumulation tests use biocon-centration factors (BCF) to predict concentrations of hydrophobic contaminants in organisms. The BCF is the ratio of the average concentration of test chemical accumulated in the tissue of the test organism (under steady state conditions) to the average measured concentration in the water.

Freshwater tests and saltwater tests have different standard methods, especially as set by the regulatory agencies. However, these tests generally include a control (negative and/or positive), a geometric dilution series or other appropriate logarithmic dilution series, test chambers and equal numbers of replicates, and a test organism. Exact exposure time and test duration will depend on type of test (acute vs. chronic) and organism type. Temperature, water quality parameters and light will depend on regulator requirements and organism type.

Effluent toxicity tests are tests conducted under the Clean Water Act, National Pollutant Discharge Elimination System (NPDES) permit program and are used by dischargers of contaminated effluent to monitor the quality of effluent into receiving waters. Acute Effluent Toxicity Tests are used to monitor the quality of industrial effluent monthly using acute toxicity tests. Effluent is used to perform static-acute multi concentration toxicity tests with *Ceriodaphnia dubia* and *Pimephales promelas*. The test organisms are exposed for 48 hours under static conditions with five concentrations of the effluent. Short-term Chronic Effluent Toxicity Tests are used to monitor the quality of municipal wastewater treatment plants effluent quarterly using short-term chronic toxicity tests. The goal of this test is to ensure that the wastewater is not chronically toxic. The major deviation in the short-term chronic effluent toxicity tests and the acute effluent toxicity tests is that the short-term chronic test lasts for seven days and the acute test lasts for 48 hours.

Sediment Tests

At some point most chemicals originating from both anthropogenic and natural sources accumulate in sediment. For this reason, sediment toxicity can play a major role in the adverse biological effects seen in aquatic organisms, especially those inhabiting benthic habitats. A recommended approach for sediment testing is to apply the Sediment Quality Triad (SQT) which involves simultaneously examining sediment chemistry, toxicity, and field alterations so that more complete information can be gathered. Collection, handling, and storage of sediment can have an effect on bioavailability and for this reason standard methods have been developed to suit this purpose.

Toxicological Effects

Toxicity can be broken down into two broad categories of direct and indirect toxicity. Direct toxicity results from a toxicant acting at the site of action in or on the organism. Indirect toxicity occurs with a change in the physical, chemical, or biological environment.

Lethality is most common effect used in toxicology and used as an endpoint for acute toxicity tests. While conducting chronic toxicity tests sublethal effects are endpoints that are looked at. These endpoints include behavioral, physiological, biochemical, histological changes.

There are a number of effects that occur when an organism is simultaneously exposed to two or more toxicants. These effects include additive effects, synergistic effects, potentiation effects, and antagonistic effects. An additive effect occurs when combined effect is equal to a combination or sum of the individual effects. A synergistic effect occurs when the combination of effects is much greater than the two individual effects added together. Potentiation is an effect that occurs when an individual chemical has no effect is added to a toxicant and the combination has a greater effect than just the toxicant alone. Finally, an antagonistic effect occurs when a combination of chemicals has less of an effect than the sum of their individual effects.

Important Aquatic Toxicology Resources

- American Society for Testing and Materials (ASTM International) – A consensus organization, representing 135 countries, that develops and delivers international voluntary standard methods for aquatic toxicity testing.

- Standard Methods for the Examination of Water and Wastewater – A compilation of techniques for the examination of water, jointly published by the American Public Health Association (APHA), the American Water Works Association (AWWA), and the Water Pollution Control Federation (WPCF).

- Ecotox – A database maintained by the U.S. Environmental Protection Agency (EPA) that offers single chemical toxicity information for both aquatic and terrestrial purposes.

- Society of Environmental Toxicology and Chemistry (SETAC) – A nonprofit, worldwide society working to promote scientific research to further our understanding of environmental stressors, environmental education, and the use of science in environmental policy.

- United States Environmental Protection Agency (EPA) – A federal agency working to protect human and environmental health. Among many other functions, the U.S. EPA produces guidance manuals outlining aquatic toxicity test procedures.

- Organisation for Economic Co-operation and Development (OECD) – A forum for governments to work together to promote policies for the betterment of people's social and economic well-being around the world. One way in which they accomplish this is through the development of aquatic toxicity test guidelines.

- Environment Canada (EC) – A diverse organization working to protect Canada's water resources and the natural environment through the coordination of environmental policies and programs with the federal government.

Terminology

- Median Lethal Concentration (LC50) – The chemical concentration that is expected to kill 50% of a group of organisms.

- Median Effective Concentration (EC50) – The chemical concentration that is expected to have one or more specified effects in 50% of a group of organisms.

- Critical Body Residue (CBR) – An approach that routinely examines whole-body chemical concentrations of an exposed organism that is associated with an adverse biological response.

- Baseline toxicity – Refers to narcosis which is a depression in biological activity due to toxicants being present in the organism.

- Biomagnification – The process by which the concentration of a chemical in the tissues of an organism increases as it passes through several levels in the food web.

- Lowest Observed Effect Concentration (LOEC) – The lowest test concentration that has a statistically significant effect over a specified exposure time.

- No Observed Effect Concentration (NOEC) – The highest test concentration for which no effect is observed relative to a control over a specified exposure time.

- Maximum Acceptable Toxicant Concentration (MATC) – An estimated value that represents the highest "no-effect" concentration of a specific substance within the range including the NOEC and LOEC.

- Application Factor (AF) – An empirically derived "safe" concentration of a chemical.

- Biomonitoring – The consistent use of living organisms to analyze environmental changes over time.

- Effluent – Liquid, industrial discharge that usually contain varying chemical toxicants.

- Quantitative Structure-Activity Relationship (QSAR) – A method of modeling the relationship between biological activity and the structure of organic chemicals.

- Mode of Action – A set of common behavioral or physiological signs that represent a type of adverse response.

- Mechanism of Action – The detailed events that take place at the molecular level during an adverse biological response.

- KOW – The octanol-water partition coefficient which represents the ratio of the concentration of octanol to the concentration of chemical in the water.

- Bioconcentration Factor (BCF) – The ratio of the average chemical concentration in the tissues of the organism under steady-state conditions to the average chemical concentration measured in the water to which the organisms are exposed.

Significance to Regulatory World

In the United States aquatic toxicology plays an important role in the NPDES wastewater permit program. In addition to analytical testing for known pollutants, aquatic, whole effluent toxicity tests have been standardized and are performed routinely as a tool for evaluating the potential harmful effects of effluents discharged into surface waters.

For the Clean Water Act under United States Environmental Protection Agency there are water quality criteria and water quality standards derived from aquatic toxicity tests.

Sediment Quality Guidelines

While sediment quality guidelines are not meant for regulation, they provide a way to rank and compare sediment quality developed by National Oceanic and Atmospheric Administration(-NOAA). These sediment quality guidelines are summarized in NOAA's Screening Quick Reference Tables (SQuiRT) for many different chemicals.

References

- Rand, Gary M.; Petrocelli, Sam R. (1985). Fundamentals of aquatic toxicology: Methods and applications. Washington: Hemisphere Publishing. ISBN 0-89116-382-4.

- "About the Organisation for Economic Co-operation and Development (OECD)" Organisation for Economic Co-operation and Development, Retrieved 2012-06-07

9

Laws Relating to Water Safety and Quality Management

The plan to ensure the safety of drinking water through the use of a comprehensive risk assessment and risk management approach is a water safety plan. The Safe Drinking Water Act (SDWA) is a law that is intended to ensure safe drinking water for the public. The topics discussed in the chapter are of great importance to broaden the existing knowledge on water safety and quality management.

Safe Drinking Water Act

The Safe Drinking Water Act (SDWA) is the principal federal law in the United States intended to ensure safe drinking water for the public. Pursuant to the act, the Environmental Protection Agency (EPA) is required to set standards for drinking water quality and oversee all states, localities, and water suppliers who implement these standards.

SDWA applies to every public water system (PWS) in the United States. There are currently about 155,000 public water systems providing water to almost all Americans at some time in their lives. The Act does not cover private wells.

The SDWA does not apply to bottled water. Bottled water is regulated by the Food and Drug Administration (FDA) under the Federal Food, Drug, and Cosmetic Act.

National Primary Drinking Water Regulations

The SDWA requires EPA to establish *National Primary Drinking Water Regulations* (NPDWRs) for contaminants that may cause adverse public health effects.

The regulations include both mandatory levels (Maximum Contaminant Levels, or MCLs) and nonenforceable health goals (Maximum Contaminant Level Goals, or MCLGs) for each included contaminant. MCLs have additional significance because they can be used under the Superfund law as "Applicable or Relevant and Appropriate Requirements" in cleanups of contaminated sites on the National Priorities List.

Federal drinking water standards are organized into six groups:

- Microorganisms
- Disinfectants
- Disinfection Byproducts
- Inorganic Chemicals

- Organic Chemicals
- Radionuclides.

Microorganisms

EPA has issued standards for *Cryptosporidium, Giardia lamblia, Legionella,* coliform bacteria and enteric viruses. EPA also requires two microorganism-related tests to indicate water quality: plate count and turbidity.

Disinfectants

EPA has issued standards for chlorine, chloramine and chlorine dioxide.

Disinfection by-products

EPA has issued standards for bromate, chlorite, haloacetic acids and trihalomethanes.

Inorganic Chemicals

EPA has issued standards for antimony, arsenic, asbestos, barium, beryllium, cadmium, chromium, copper, cyanide, fluoride, lead, mercury, nitrate, nitrite, selenium and thallium.

"Lead Free" Plumbing Requirements

The 1986 amendments require EPA to set standards limiting the concentration of lead in public water systems, and defines "lead free" pipes as:

(1) solders and flux containing not more than 0.2 percent lead;

(2) pipes and pipe fittings containing not more than 8.0 percent lead; and

(3) plumbing fittings and fixtures as defined in industry-developed voluntary standards (issued no later than August 6, 1997), or standards developed by EPA in lieu of voluntary standards.

EPA issued an initial lead and copper regulation in 1991 and last revised the regulation in 2007.

Congress tightened the definition of "lead free" plumbing in a 2011 amendment to the Act.

Organic Chemicals

EPA has issued standards for 53 organic compounds, including benzene, dioxin (2,3,7,8-TCDD), PCBs, styrene, toluene, vinyl chloride and several pesticides.

Radionuclides

EPA has issued standards for alpha particles, beta particles and photon emitters, radium and uranium.

Future standards

Non-community Water Systems

Future NPDWR standards will apply to non-transient non-community water systems because of concern for the long-term exposure of a stable population. It is important to note that EPA's decision to apply future NPDWRs to non-transient non-community water systems may have a significant impact on Department of Energy facilities that operate their own drinking water systems.

Unregulated Contaminants

The SDWA requires EPA to identify and list unregulated contaminants which may require regulation. The Agency must publish this list, called the "Contaminant Candidate List," every five years. EPA is required to decide whether to regulate at least five or more listed contaminants. EPA uses this list to prioritize research and data collection efforts, which support the regulatory determination process.

Monitoring, Compliance and Enforcement

Public water systems are required to regularly monitor their water for contaminants. Water samples must be analyzed using EPA-approved testing methods, by laboratories that are certified by EPA or a state agency.

A PWS must notify its customers when it violates drinking water regulations or is providing drinking water that may pose a health risk. Such notifications are provided either immediately, as soon as possible (but within 30 days of the violation) or annually, depending on the health risk associated with the violation. Community water systems—those systems that serve the same people throughout the year—must provide an annual "Consumer Confidence Report" to customers. The report identifies contaminants, if any, in the drinking water and explains the potential health impacts.

Oversight of public water systems is managed by "primacy" agencies, which are either state government agencies, Indian tribes or EPA regional offices. All state and territories, except Wyoming and the District of Columbia, have received primacy approval from EPA, to supervise the PWS in their respective jurisdictions. A PWS is required to submit periodic monitoring reports to its primacy agency. Violations of SDWA requirements are enforced initially through a primacy agency's notification to the PWS, and if necessary following up with formal orders and fines.

Related Programs

Airline Water Supplies

In 2004, EPA tested drinking water quality on commercial aircraft and found that 15 percent of tested aircraft water systems tested positive for total coliform bacteria. EPA published a final regulation for aircraft public water systems in 2009. The regulation requires air carriers operating in the U.S. to conduct coliform sampling, management practices, corrective action, public notification, operator training, and reporting and recordkeeping. An airline with a non-complying aircraft must restrict public access to the on-board water system for a specified period.

Underground Injection Control (UIC) Program

The 1974 act authorized EPA to regulate injection wells in order to protect underground sources of drinking water. Congress amended the SDWA in 2005 to exclude hydraulic fracturing, an industrial process for recovering oil and natural gas, from coverage under the UIC program, except where diesel fuels are used. This exclusion has been called the "Halliburton Loophole". Halliburton is the world's largest provider of hydraulic fracturing services.

Whistleblower Protection

The SDWA includes a whistleblower protection provision. Employees in the US who believe they were fired or suffered another adverse action related to enforcement of this law have 30 days to file a written complaint with the Occupational Safety and Health Administration.

History

Prelude

Prior to the SDWA there were few national enforceable requirements for drinking water. Improvements in testing were allowing the detection of smaller concentrations of contaminant and allowing more tests to be run. Many states had drinking water regulations prior to adoption of the federal SDWA.

1974 Act

The Safe Drinking Water Act was one of several pieces of environmental legislation in the 1970s. Discovery of organic contamination in public drinking water and the lack of enforceable, national standards persuaded Congress to take action.

1986 Amendments

The 1986 SDWA amendments required EPA to apply future NPDWRs to both community and non-transient non-community water systems when it evaluated and revised current regulations. The first case in which this was applied was the "Phase I" final rule, published on July 8, 1987. At that time NPDWRs were promulgated for certain synthetic volatile organic compounds and applied to non-transient non-community water systems as well as community water systems. This rulemaking also clarified that non-transient non-community water systems were not subject to MCLs that were promulgated before July 8, 1987. The 1986 amendments were signed into law by President Ronald Reagan on June 19, 1986.

In addition to requiring more contaminants to be regulated, the 1986 amendments included:

- Well head protection

- New monitoring for certain substances

- Filtration for certain surface water systems

- Disinfection for certain groundwater systems

- Restriction on lead in solder and plumbing

- More enforcement powers.

1996 SDWA Amendments

In 1996, Congress amended the Safe Drinking Water Act to emphasize sound science and risk-based standard setting, small water supply system flexibility and technical assistance, community-empowered source water assessment and protection, public right-to-know, and water system infrastructure assistance through a multibillion-dollar state revolving loan fund. The amendments were signed into law by President Bill Clinton on August 6, 1996.

Main Points of the 1996 Amendments

1. Consumer Confidence Reports: All community water systems must prepare and distribute annual reports about the water they provide, including information on detected contaminants, possible health effects, and the water's source.

2. Cost-Benefit Analysis: EPA must conduct a thorough cost-benefit analysis for every new standard to determine whether the benefits of a drinking water standard justify the costs.

3. Drinking Water State Revolving Fund. States can use this fund to help water systems make infrastructure or management improvements or to help systems assess and protect their source water.

4. Microbial Contaminants and Disinfection Byproducts: EPA is required to strengthen protection for microbial contaminants, including *cryptosporidium,* while strengthening control over the byproducts of chemical disinfection. EPA promulgated the *Stage 1 Disinfectants and Disinfection Byproducts Rule* and the *Interim Enhanced Surface Water Treatment Rule* to address these risks.

5. Operator Certification: Water system operators must be certified to ensure that systems are operated safely. EPA issued guidelines in 1999 specifying minimum standards for the certification and recertification of the operators of community and non-transient, noncommunity water systems. These guidelines apply to state operator certification programs. All states are currently implementing EPA-approved operator certification programs.

6. Public Information and Consultation: SDWA emphasizes that consumers have a right to know what is in their drinking water, where it comes from, how it is treated, and how to help protect it. EPA distributes public information materials (through its Drinking Water Hotline, Safewater web site, and Resource Center) and holds public meetings, working with states, tribes, water systems, and environmental and civic groups, to encourage public involvement.

7. Small Water Systems: Small water systems are given special consideration and resources under SDWA, to make sure they have the managerial, financial, and technical ability to comply with drinking water standards.

2005 Amendment

Through the Energy Policy Act of 2005, the Safe Drinking Water Act was amended to exclude the underground injection of any fluids or propping agents other than diesel fuels used in hydraulic fracturing operations from being considered as "underground injections" for the purposes of the law.

2011 Amendment

Congress passed the *Reduction of Lead in Drinking Water Act* in 2011. This amendment, effective in 2014, tightened the definition of "lead-free" plumbing fixtures and fittings.

2015 Amendments

The *Drinking Water Protection Act* was enacted on August 7, 2015. It required EPA to submit to Congress a strategic plan for assessing and managing risks associated with algal toxins in drinking water provided by public water systems. EPA submitted the plan to Congress in November 2015.

The *Grassroots Rural and Small Community Water Systems Assistance Act* was signed by President Barack Obama on December 11, 2015. The amendment provides technical assistance to small public water systems, to help them comply with National Primary Drinking Water Regulations.

Clean Water Act

The Clean Water Act (CWA) is the primary federal law in the United States governing water pollution. Its objective is to restore and maintain the chemical, physical, and biological integrity of the nation's waters by preventing point and nonpoint pollution sources, providing assistance to publicly owned treatment works for the improvement of wastewater treatment, and maintaining the integrity of wetlands. It is one of the United States' first and most influential modern environmental laws. As with many other major U.S. federal environmental statutes, it is administered by the U.S. Environmental Protection Agency (EPA), in coordination with state governments. Its implementing regulations are codified at 40 C.F.R. Subchapters D, N, and O (Parts 100-140, 401-471, and 501-503).

Technically, the name of the law is the Federal Water Pollution Control Act. The first FWPCA was enacted in 1948, but took on its modern form when completely rewritten in 1972 in an act entitled the Federal Water Pollution Control Act Amendments of 1972. Major changes have subsequently been introduced via amendatory legislation including the Clean Water Act of 1977 and the Water Quality Act of 1987.

The Clean Water Act does not directly address groundwater contamination. Groundwater protection provisions are included in the Safe Drinking Water Act, Resource Conservation and Recovery Act, and the Superfund act.

Waters Protected under the CWA

All waters with a "significant nexus" to "navigable waters" are covered under the CWA; however, the phrase "significant nexus" remains open to judicial interpretation and considerable controversy. The 1972 statute frequently uses the term "navigable waters," but also defines the term as "waters of the United States, including the territorial seas." Some regulations interpreting the 1972 law have included water features such as intermittent streams, playa lakes, prairie potholes, sloughs and wetlands as "waters of the United States." In the 2006 case *Rapanos v. United States,* a plurality of the Supreme Court held that the term "waters of the United States":

> ...includes only those relatively permanent, standing or continuously flowing bodies of water "forming geographic features" that are described in ordinary parlance as "streams[,] ... oceans, rivers, [and] lakes."

Pollution Control Strategy in the CWA

Point Sources

The 1972 act introduced the National Pollutant Discharge Elimination System (NPDES), which is a permit system for regulating point sources of pollution. Point sources include:

- industrial facilities (including manufacturing, mining, oil and gas extraction, and service industries).

- municipal governments and other government facilities (such as military bases), and

- some agricultural facilities, such as animal feedlots.

Point sources may not discharge pollutants to surface waters without a permit from the National Pollutant Discharge Elimination System (NPDES). This system is managed by the United States Environmental Protection Agency (EPA) in partnership with state environmental agencies. EPA has authorized 46 states to issue permits directly to the discharging facilities. The CWA also allows tribes to issue permits, but no tribes have been authorized by EPA. In the remaining states and territories, the permits are issued by an EPA regional office.

In previous legislation, Congress had authorized states to develop water quality standards, which would limit discharges from facilities based on the characteristics of individual water bodies. However, these standards were only to be developed for interstate waters, and the science to support this process (i.e. data, methodology) was in the early stages of development. This system was not effective and there was no permit system in place to enforce the requirements. In the 1972 CWA Congress added the permit system and a requirement for technology-based effluent limitations.

Technology-based Standards

The 1972 CWA created a new requirement for technology-based standards for point source discharges. EPA develops these standards for categories of dischargers, based on the performance of pollution control technologies without regard to the conditions of a particular receiving water body. The intent of Congress was to create a "level playing field" by establishing a basic national discharge standard for all facilities within a category, using a "Best Available Technology." The standard becomes the minimum regulatory requirement in a permit. If the national standard is not sufficiently protective at a particular location, then water quality standards may be employed.

Water Quality Standards

The 1972 act authorized continued use of the water quality-based approach, but in coordination with the technology-based standards. After application of technology-based standards to a permit, if water quality is still impaired for the particular water body, then the permit agency (state or EPA) may add water quality-based limitations to that permit. The additional limitations are to be more stringent than the technology-based limitations and would require the permittee to install additional controls. Water quality standards consist of four basic elements: 1) Designated uses; 2) Water quality criteria; 3) Antidegradation policy and 4) General policies.

Designated Uses

According to water quality standard regulations, states and Indian tribes are required to specify appropriate water uses. Identification of appropriate water uses takes into consideration the usage and value of public water supply, protection of fish, wildlife, recreational waters, agricultural, industrial and navigational water ways. Suitability of a water body is examined by states and tribes for usages based on physical, chemical, and biological characteristics. States and Indian tribes also examine geographical settings, scenic qualities and economic considerations to determine fitness of designated uses for a water bodies. If these standards indicate designated uses to be less than those presently attained, states or tribes are required to revise standards to reflect the uses actually being attained. For any body of water with designated uses that do not include "fishable/swimmable" target use that is identified in section 101(a)(2) of CWA, a Use Attainability Analysis must be conducted. Every three years, such bodies of water must be reexamined in order to verify if new information is available that demand a revision of the standard. If new information is available that specify "fishable/swimmable" uses can be attained, then the use must be designated.

Water Quality Criteria

States and tribes protect designated areas by adopting water quality criteria that allow them to adopt the criteria that EPA publishes under §304(a) of the CWA, modify the §304(a) criteria to reflect site-specific conditions or adopt criteria based on other scientifically defensible methods. Water quality criteria can be numeric criteria that toxicity causes are known for protection against pollutants. A narrative criterion is water quality criteria which serves as basis for limiting toxicity of waste discharge to aquatic species. A biological criterion is based on aquatic community which describes the number and types of species in a water body. A nutrient criterion solely protects against nutrient over enrichment; and a sediment criterion describes conditions of contaminated and uncontaminated sediments in order to avoid undesirable effects.

Anti-degradation Policy

Water quality standards consist of an anti-degradation policy that requires states and tribes to establish a three-tiered anti-degradation program. Anti-degradation procedures identify steps and questions that need to be addressed when specific activities affect water quality. Tier 1 is applicable to all surface waters. It maintains and protects current uses and water quality conditions to support existing uses. Current uses are identified by showing that fishing, swimming, and other water uses have occurred and are suitable since November 28, 1975. Tier 2 maintains and protects water bodies with existing conditions that are better to support CWA 101(a)(2) "fishable/swimmable" uses. Tier 3 maintains and protects water quality in outstanding national resource waters (ONR-Ws), which are the highest quality waters in the US with ecological significance.

General Policies

States and Indian tribes adopt general policies pertaining to water quality standards that are subject to review and approval by the EPA. These provision regarding water quality standards include mixing zones, variance, and low flow policies. Mixing zone policy is defined area surrounding a point source discharge where sewage is diluted by water. Methodology of mixing zone procedure determines the location, size, shape and quality of mixing zones. Variance policy temporarily relax water quality standard and are alternatives to removing a designated use. States and tribes may include variance as part of their water quality standard. Variance is subject to public review every three years and warrant development towards improvement of water quality. Low Flow policy pertains to states and tribes water quality standards that identify procedures applied to determining critical low flow conditions.

Nonpoint Sources

Congress exempted some water pollution sources from the point source definition in the 1972 CWA, and was unclear on the status of some other sources. These sources were therefore considered to be nonpoint sources that were not subject to the permit program.

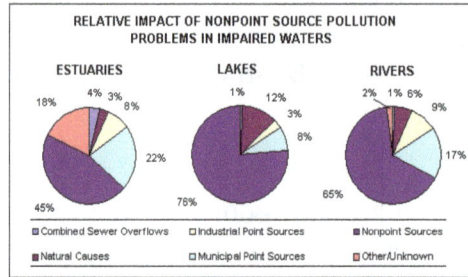

Nonpoint source pollutants, such as sediments, nutrients, pesticides, herbicides, fertilizers,animal wastes and other substances that enter our water supply as components of runoff and ground water, have increased in relative significance and accounts for more than 50 percent of the pollution in U.S. waters.

Agricultural stormwater discharges and irrigation return flows were specifically exempted from permit requirements. Congress, however, provided support for research, technical and financial assistance programs at the U.S. Department of Agriculture to improve runoff management prac-tices on farms.

Stormwater runoff from industrial sources, municipal storm drains, and other sources were not specifically addressed in the 1972 law. EPA declined to include urban runoff and industrial stormwater discharges in the NPDES program and consequently was sued by an environmental group. The courts ruled that stormwater discharges must be covered by the permit program.

A growing body of water research during the late 1970s and 1980s indicated that stormwater runoff was a significant cause of water quality impairment in many parts of the U.S. In the early 1980s EPA conducted the Nationwide Urban Runoff Program (NURP) to document the extent of the urban stormwater problem. The agency began to develop regulations for stormwater permit coverage, but encountered resistance from industry and municipalities, and there were additional rounds of litigation. This litigation was pending when Congress considered further amendments to the Act in 1986.

In the Water Quality Act of 1987 (1987 WQA) Congress responded to the stormwater problem by requiring that industrial stormwater dischargers and municipal separate storm sewer systems (often called "MS4") obtain NPDES permits, by specific deadlines. The permit exemption for agricultural discharges continued, but Congress created a nonpoint source pollution demonstration grant program at EPA to expand the research and development of nonpoint controls and management practices.

To combat nonpoint source pollution, EPA initiated numerous programs and grants to aid the public in improving their local water quality. These programs are described at an EPA website, Watershed Central.

Financing of Pollution Controls

Congress created a major public works financing program for municipal sewage treatment in the 1972 CWA. A system of grants for construction of municipal sewage treatment plants was authorized and funded in Title II. In the initial program the federal portion of each grant was up to 75 percent of a facility's capital cost, with the remainder financed by the state. In subsequent amendments Congress reduced the federal proportion of the grants and in the 1987 WQA transitioned to a revolving loan program in Title VI. Industrial and other private facilities are required to finance their own treatment improvements on the "polluter pays" principle.

Major Statutory Provisions

This Act has six titles.

Title I - Research and Related Programs

Title I includes a *Declaration of Goals and Policy* and various grant authorizations for research programs and pollution control programs. Some of the programs authorized by the 1972 law are ongoing (e.g. section 104 research programs, section 106 pollution control programs, section 117 Chesapeake Bay Program) while other programs no longer receive funds from Congress and have been discontinued.

Title II - Grants for Construction of Treatment Works

To assist municipalities in creating or expanding sewage treatment plants, also known as publicly owned treatment works (POTW), Title II established a system of construction grants. This was replaced by the Clean Water State Revolving Fund in the 1987 WQA.

Title III - Standards and enforcement

Discharge Permits Required

Section 301 of the Act prohibits discharges to waters of the U.S. except with a permit.

Technology-Based Standards Program

Under the 1972 act EPA began to issue technology-based standards for municipal and industrial sources.

- Municipal sewage treatment plants (POTW) are required to meet secondary treatment standards.

- Effluent guidelines (for existing sources) and New Source Performance Standards (NSPS) are issued for categories of industrial facilities discharging directly to surface waters.

- Categorical Pretreatment Standards are issued to industrial users (also called "indirect dischargers") contributing wastes to POTW. These standards are developed in conjunction with the effluent guidelines program. As with effluent guidelines and NSPS, pretreatment standards consists of Pretreatment Standards for Existing Sources (PSES) and Pretreatment Standards for New Sources (PSNS). There are 27 categories with pretreatment standards as of 2011.

To date, the effluent guidelines and categorical pretreatment standards regulations have been published for 56 categories and apply to between 35,000 and 45,000 facilities that discharge directly to the nation's waters. These regulations are responsible for preventing the discharge of almost 700 billion pounds of pollutants each year. EPA has updated some categories since their initial promulgation and has added new categories.

The secondary treatment standards for POTWs and the effluent guidelines are implemented through NPDES permits. The categorical pretreatment standards are typically implemented by POTWs through permits that they issue to their industrial users.

Water Quality Standards Program

Water quality standards (WQS) are risk-based requirements which set site-specific allowable pollutant levels for individual water bodies, such as rivers, lakes, streams and wetlands. States set WQS by designating uses for the water body (e.g., recreation, water supply, aquatic life, agriculture) and applying water quality criteria (numeric pollutant concentrations and narrative requirements) to protect the designated uses. An antidegradation policy is also issued by each state to maintain and protect existing uses and high quality waters.

Water bodies that do not meet applicable water quality standards with technology-based controls alone are placed on the section 303(d) list of water bodies not meeting standards. Water bodies on the 303(d) list require development of a Total Maximum Daily Load (TMDL). A TMDL is a calculation of the maximum amount of a pollutant that a water body can receive and still meet WQS. The TMDL is determined after study of the specific properties of the water body and the pollutant sources that contribute to the non-compliant status. Generally, the TMDL determines load based on a Waste Load Allocation (WLA), Load Allocation (LA), and Margin of Safety (MOS) Once the TMDL assessment is completed and the maximum pollutant loading capacity defined, an implementation plan is developed that outlines the measures needed to reduce pollutant loading to the non-compliant water body, and bring it into compliance. Over 60,000 TMDLs are proposed or in development for U.S. waters in the next decade and a half.

Following the issuance of a TMDL for a water body, implementation of the requirements involves modification to NPDES permits for facilities discharging to the water body to meet the WLA allocated to the water body.

As of 2007, approximately half of the rivers, lakes, and bays under EPA oversight were not safe enough for fishing and swimming. The development of WQS and TMDL is a complex process, both scientifically and legally, and it is a resource-intensive process for state agencies.

National Water Quality Inventory

The primary mode of informing the quality of water of rivers, lakes, streams, ponds, estuaries, coastal waters and wetlands of the U.S. is through the *National Water Quality Inventory Report*. Water quality assessments are conducted pursuant to water quality standards adopted by states and other jurisdictions (territories, interstate commissions and tribes). The report is conveyed to Congress as a means to inform Congress and the public of compliance with quality standards established by states, territories and tribes. The assessments identify water quality problems within the states and jurisdictions, list the impaired and threatened water bodies, and identify non-point sources that contribute to poor water quality. Every two years states must submit reports that describe water quality conditions to EPA with a complete inquiry of social and economic costs and benefits of achieving goals of the Act. The report is organized into two major sections; Section 1 shows national assessment of each type of water body, with causes and sources identified. Section 2 summarizes recommendations on improvement of water resource management.

Enforcement

Under section 309, EPA can issue administrative orders against violators, and seek civil or criminal penalties when necessary.

- For a first offense of criminal negligence, the minimum fine is $2,500, with a maximum of $25,000 fine per day of violation. A violator may also receive up to a year in jail. On a second offense, a maximum fine of $50,000 per day may be issued.

- For a knowing endangerment violation, i.e. placing another person in imminent danger of death or serious bodily injury, a fine may be issued up to $250,000 and/or imprisonment up to 15 years for an individual, or up to $1,000,000 for an organization.

States that are authorized by EPA to administer the NPDES program must have authority to enforce permit requirements under their respective state laws.

Federal Facilities

Military bases, national parks and other federal facilities must comply with CWA provisions.

Thermal Pollution

Section 316 requires standards for thermal pollution discharges, as well as standards for cooling water intake structures. These standards are applicable to power plants and other industrial facilities.

Nonpoint Source Management Program

The 1987 amendments created the Nonpoint Source Management Program under CWA section 319. This program provides grants to states, territories and Indian tribes to support demonstration projects, technology transfer, education, training, technical assistance and related activities designed to reduce nonpoint source pollution. Grant funding for the program averaged $210 million annually for Fiscal Years 2004 through 2008.

Title IV - Permits and Licenses

State certification of Compliance

States are required to certify that discharges authorized by federal permits will not violate the state's water quality standards.

NPDES Permits for Point Sources

The NPDES permits program is authorized by CWA section 402. The initial permits issued in the 1970s and early 1980s focused on POTWs and industrial wastewater—typically "process" wastewater and cooling water where applicable, and in some cases, industrial stormwater. The 1987 WQA expanded the program to cover stormwater discharges explicitly, both from municipal separate storm sewer systems (MS4) and industrial sources. The MS4 NPDES permits require regulated municipalities to use Best Management Practices to reduce pollutants to the "Maximum Extent Practicable."

Non-stormwater permits typically include numeric effluent limitations for specific pollutants. A numeric limitation quantifies the maximum pollutant load or concentration allowed in the discharge, e.g., 30 mg/L of biochemical oxygen demand. Exceeding a numeric limitation constitutes a violation of the permit, and the discharger is subject to fines as laid out in section 309. Facilities must periodically monitor their effluent (i.e., collect and analyze wastewater samples), and submit Discharge Monitoring Reports to the appropriate agency, to demonstrate compliance. Stormwater permits typically require facilities to prepare a Stormwater Pollution Prevention Plan and implement best management practices, but do not specify numeric effluent limits and may not include regular monitoring requirements. Some permits cover both stormwater and non-stormwater discharges. NPDES permits must be reissued every five years. Permit agencies (EPA, states, tribes) must provide notice to the public of pending permits and provide an opportunity for public comment.

As of 2001, over 400,000 facilities were subject to NPDES permit requirements. This number includes permanent facilities such as municipal (POTW, MS4) and industrial plants, and construction sites, which are temporary stormwater dischargers.

Dredge and Fill Exemptions

After passage of the CWA in 1972, a controversy arose as to its application to agriculture and certain other activities. The Act was interpreted by some to place restrictions on virtually all placement of dredged materials in wetlands and other waters of the United States, raising concern that the federal government was about to place all agricultural activities under the jurisdiction of the U.S. Army Corps of Engineers (USACE). For opponents of the Act, section 404 had, as a result of this concern, become a symbol of dramatic over-regulation. When Congress considered the 1977 CWA Amendments, a significant issue was to ensure that certain agricultural activities and other selected activities, could continue without the government's supervision—in other words, completely outside the regulatory or permit jurisdiction of any federal agency.

The 1977 amendments included a set of six section 404 exemptions. For example, totally new activities such as construction of farm roads, Sec. 1344(f)(1)(E), construction of farm or stock ponds or irrigation ditches, and minor agricultural drainage, Sec. 1344(f)(1)(A), all are exempted by Statute. Section 1344(f)(1)(C), which exempts discharge of dredged material "for the purpose of. . . the maintenance of drainage ditches." All of these exemptions were envisioned to be self-executing, that is not technically requiring an administrative no-jurisdiction determination. One such example was the maintenance of agricultural drainage ditches. Throughout the hearing process, Congressmen of every environmental persuasion repeatedly stated that the over $5 Billion invested in drainage facilities could be maintained without government regulation of any kind. Senator Edmund Muskie, for example, explained that exempt activities such as agricultural drainage would be entirely unregulated. Other exemptions were granted as well, including exemptions for normal farming activities.

Importance of no-jurisdiction Determinations

Although Congress envisioned a set of self-executing exemptions, it has become common for landowners to seek no-jurisdiction determinations from the USACE. A landowner who intends to make substantial investments in acquisition or improvement of land might lawfully proceed with exempt activity, a permit not being required. The problem is that if the landowner's assumptions were incorrect and the activity later determined not to be exempt, the USACE will issue a cease and desist order. Obtaining an advanced ruling provides some level of comfort that the activities will have been deemed conducted in good faith.

Recapture of Exemptions

Because some of the six exemptions involved new activities, such as minor drainage and silviculture (the clearing of forests by the timber industry), Congress recognized the need to impose some limitations on exemptions. Consequently, Congress placed the so-called recapture clause limitation on these new project exemptions. Under section 404(f)(2), such new projects would be deprived of their exemption if all of the following three characteristics could be shown:

1. A discharge of dredge or fill material in the navigable waters of the United States;

2. The discharge is incidental to an activity having as its purpose the bringing of an area of navigable waters into a use to which it was not previously subject, and

3. Where the flow or circulation of navigable waters may be impaired or the reach of such waters may be reduced.

To remove the exemption, all of these requirements must be fulfilled—the discharge, the project purpose of bringing an area into a use to which it was not previously subject, and the impairment or reduction of navigable waters.

Dredge and Fill Permits (Wetlands, Lakes, Streams, Rivers, and other Waters of the U.S.)

Under sections 301 and 502 of the Clean Water Act, any discharge of dredged or fill materials into "waters of the United States," including wetlands, is forbidden unless authorized by a permit issued by the USACE pursuant to section 404. Essentially, all discharges of fill or dredged material affecting the bottom elevation of a jurisdictional water of the U.S. require a Department of the Army (DA) permit from USACE. These permits are an essential part of protecting streams and wetlands, which are often filled by land developers. Wetlands are vital to the ecosystem in filtering streams and rivers and providing habitat for wildlife.

Mountaintop removal mining requires a section 404 permit when soil and rock from the mining operation is placed in streams and wetlands (commonly called a "valley fill"). Pollutant discharges from valley fills to streams also requires an NPDES permit.

There are two main types of wetlands permits: general permits and individual permits. General permits change periodically and cover broad categories of activities, and require the permittee to comply with all stated conditions. General permits (such as the Nationwide Permits) are issued for fill activities that will result in minimal adverse effects to the environment. Individual permits are utilized for actions that are not addressed by a general permit, or that do not meet the conditions of a General Permit. In addition, individual permits typically require more analysis than do the general permits, and usually require much more time to prepare the application and to process the permit.

When the USACE processes an application for an Individual Permit, it must publish/issue a public notice describing the proposed action described in the permit application. Although the Corps District Engineer makes the decision to grant a permit, the USEPA Administrator may veto a permit if it is not reasonable. Before making such a decision, however, USEPA must consult with the USACE. A DA permit typically expires after five years.

POTW Biosolids Management Program

The 1987 WQA created a program for management of biosolids (sludge) generated by POTWs. The Act instructed EPA to develop guidelines for usage and disposal of sewage sludge or biosolids. The EPA regulations: (1) Identify uses for sewage sludge, including disposal; (2) Specify factors to be taken into account in determining the measures and practices applicable to each such use or disposal (including publication of information on costs); and (3) Identify concentrations of pollutants which interfere with each such use or disposal. EPA created an Intra-Agency Sludge Task Force to aid in developing comprehensive sludge regulations that are designed to do the following: (1) Conduct a multimedia examination of sewage sludge management, focusing on sewage sludge generated by POTWs; and (2) develop a cohesive Agency policy on sewage sludge management, designed to guide the Agency in implementing sewage sludge regulatory and management programs.

The term *biosolids* is used to differentiate treated sewage sludge that can be beneficially recycled. Environmental advantages of sewage sludge consist of, application of sludge to land due to its soil condition properties and nutrient content. Advantages also extend to reduction in adverse health effects of incineration, decreased chemical fertilizer dependency, diminishing greenhouse gas emissions deriving from incineration and reduction in incineration fuel and energy costs. Beneficial reuse of sewage sludge is supported in EPA policies: the 1984 *Beneficial Reuse Policy* and the 1991 *Inter-agency Policy on Beneficial Use of Sewage Sludge,* with an objective to reduce volumes of waste generated. Sewage sludge contains nutrients such as nitrogen and phosphorus but also contains significant numbers of pathogens such as bacteria, viruses, protozoa and eggs of parasitic worms. Sludge also contains more than trace amounts of organic and inorganic chemicals. Benefits of reusing sewage sludge from use of organic and nutrient content in biosolids is valuable source in improving marginal lands and serving as supplements to fertilizers and soil conditioners. Extension of benefits of sludge on agriculture commodities include increase forest productivity, accelerated tree growth, re-vegetation of forest land previously devastated by natural disasters or construction activities. Also, sewage sludge use to aid growth of final vegetative cap for municipal solid waste landfills is enormously beneficial. Opposing benefits of sludge water result from high levels of pathogenic organisms that can possibly contaminate soil, water, crops, livestock, and fish. Pathogens, metals, organic chemical content and odors are cause of major health, environmental and aesthetic factors. Sludge treatment processes reduce the level of pathogens which becomes important when applying sludge to land as well as distributing and marketing it. Pollutants of sewage sludge come from domestic wastewater, discharge of industrial wastewater, municipal sewers and also from runoffs from parking lots, lawns and fields that were applied fertilizers, pesticides and insecticides.

The quality of sewage sludge is controlled under section 405(d), where limitations are set with methods of use or disposal for pollutants in sludge. EPA, under section 405(d)(3), established a containment approach to limit pollutants instead of numerical limitations. This methodology is more reasonable than numerical limitations and includes design standards, equipment standards, management practice, and operational standards or combination of these. Limits on sewage sludge quality allows treatment works that generate less contaminated pollutants and those that do not meet the sludge quality standards for use and disposal practice must clean up influent, improve sewage sludge treatment and/or select another use of disposal method. EPA has set standards for appropriate practices of use and disposal of biosolids in order to protect public health and the environment, but choice of use or disposal practices are reserved to local communities. Listed under section 405(e) of CWA, local communities are encouraged to use their sewage sludge for its beneficial properties instead of disposing it.

Standards are set for sewage sludge generated or treated by publicly owned and privately owned treatment works that treat domestic sewage and municipal wastewater. Materials flushed in household drains through sinks, toilets and tubs are referred to as domestic wastewater and include components of soaps, shampoos, human excrement, tissues, food particles, pesticides, hazardous waste, oil and grease. These domestic wastewaters are treated at the source in septic tanks, cesspools, portable toilets, or in publicly/privately owned wastewater treatment works. Alternately, municipal wastewater treatments consist of more levels of treatment that provide greater wastewater cleanup with larger amounts of sewage sludge. Primary municipal treatment remove solids that settle at the bottom, generating more than 3,000 liters of sludge per million liters of wastewater that is treated. Primary sludge water content is easily reduced by thickening or removing water and contains up to 7% solids. Secondary municipal treatment process produces sewage sludge that is generated by biological treatment processes that include activated sludge systems, trickling filters, and other attached growth systems. Microbes are used to break down and convert organic substances in wastewater to microbial residue in biological treatment processes. This process removes up to 90% of organic matter and produces sludge that contains up to 2% solids and has increased generated volumes of sludge. Methods of use and disposal of sewage sludge include the following: Application of sludge to agricultural and non-agricultural lands; sale or give-away of sludge for use in home gardens; disposal of sludge in municipal landfills, sludge-only landfills, surface disposal sites and incineration of sludge. Managing quality of sewage sludge not only involves wastewater reduction and separation of contaminated waste from non-contaminants but also pretreatment of non-domestic wastewater. Pretreatment does not thoroughly reduce pollutants level and therefore communities have to dispose rather than use sludge.

Title V - General Provisions

Citizen Suits

Any U.S. citizen may file a citizen suit against any person who has allegedly violated an effluent limitation regulation or against the EPA Administrator if the EPA Administrator failed to perform any non-discretionary act or duty required by the CWA.

Employee Protection

The CWA includes an employee ("whistleblower") protection provision. Employees in the U.S. who believe they were fired or suffered adverse action related to enforcement of the CWA may file a written complaint with the Occupational Safety and Health Administration.

Title VI - State Water Pollution Control Revolving Funds

The Clean Water State Revolving Fund (CWSRF) program was authorized by the 1987 WQA. This replaced the municipal construction grants program, which was authorized in the 1972 law under Title II. In the CWSRF, federal funds are provided to the states and Puerto Rico to capitalize their respective revolving funds, which are used to provide financial assistance (loans or grants) to local governments for wastewater treatment, nonpoint source pollution control and estuary protection.

The fund provides loans to municipalities at lower-than-market rates. As of 2009 the average rate was 2.3 percent nationwide, compared to an average market rate of 5 percent. In 2009, CWSRF assistance totaling $5.2 billion was provided to 1,971 local projects across the country.

Earlier Legislation

During the 1880s and 1890s, Congress directed USACE to prevent dumping and filling in the nation's harbors, and the program was vigorously enforced. Congress first addressed water pollution issues in the Rivers and Harbors Act of 1899, giving the Corps the authority to regulate most kinds of obstructions to navigation, including hazards resulting from effluents. Portions of this law remain in effect, including Section 13, the so-called Refuse Act. In 1910, USACE used the act to object to a proposed sewer in New York City, but a court ruled that pollution control was a matter left to the states alone. Speaking to the 1911 National Rivers and Harbors Congress, the chief of the Corps, Brigadier General William H. Bixby, suggested that modern treatment facilities and prohibitions on dumping "should either be made compulsory or at least encouraged everywhere in the United States."

Some sections of the 1899 act have been superseded by various amendments, including the 1972 CWA, while other notable legislative predecessors include:

- *Public Health Service Act of 1912* expanded the mission of the United States Public Health Service to study problems of sanitation, sewage and pollution.

- *Oil Pollution Act of 1924* prohibited the intentional discharge of fuel oil into tidal waters and provided authorization for USACE to apprehend violators. This was repealed by the 1972 CWA, reducing the Corps' role in pollution control to the discharge of dredged or fill material.

- *Federal Water Pollution Control Act* of 1948 created a comprehensive set of water quality programs that also provided some financing for state and local governments. Enforcement was limited to interstate waters. The Public Health Service provided financial and technical assistance.

- *Water Quality Act of 1965* required states to issue water quality standards for interstate waters, and authorized the newly created Federal Water Pollution Control Administration to set standards where states failed to do so.

Case Law

- *United States v. Riverside Bayview Homes, Inc.* (1985). The Supreme Court upheld the Act's coverage in regulating wetlands that intermingle with navigable waters. This ruling was revised by the 2006 *Rapanos* decision.

- *Edward Hanousek, Jr v. United States* (9th Cir. Court of Appeals, 1996; certiorari denied, 2000). In 1994, during rock removal operations, a backhoe operator accidentally struck a petroleum pipeline near the railroad tracks. The operator's mistake caused the pipeline to rupture and spill between 1,000 and 5,000 gallons of heating oil into the Skagway river. Despite not being present at the scene during operations White Pass and Yukon Route Roadmaster Edward Hanousek, Jr. and President Paul Taylor were both held responsible for the spill and convicted.

- *Solid Waste Agency of North Cook County (SWANCC) v. United States Army Corps of Engineers* (2001), possibly denying the CWA's hold in isolated intrastate waters and certainly denying the validity of the 1986 "Migratory Bird Rule."

- *S. D. Warren Co. v. Maine Bd. of Env. Protection* (2006). The Court ruled that section 401 state certification requirements apply to hydroelectric dams, which are federally licensed, where the dams cause a discharge into navigable waters.

- *Rapanos v. United States* (2006). The Supreme Court questioned federal jurisdiction as it attempted to define the Act's use of the terms "navigable waters" and "waters of the United States." The Court rejected the position of the USACE that its authority over water was essentially limitless. Though the case resulted in no binding case law, the Court suggested a narrowing of federal jurisdiction and implied the federal government needed a more substantial link between navigable federal waters and wetlands than it had been using, but held onto the "significant nexus" test.

- *National Cotton Council v. EPA* (6th Cir. Court of Appeals, 2009). Point source discharges of biological pesticides, and chemical pesticides that leave a residue, into waters of the U.S. are subject to NPDES permit requirements.

- *Army Corps of Engineers v. Hawkes Co.* 578 U.S. (2016), 8-0 ruling that a jurisdictional determination by the Army Corps of Engineers that land contains "waters of the United States" is a "final agency action", which is reviewable by the courts. This allows landowners to sue in court if the Army Corps of Engineers determines that the land contains waters of the United States (and therefore falls under the Clean Water Act).

Recent Developments

Rep. David Joyce introduced H.R. 223, the "Great Lakes Restoration Initiative Act of 2015," on January 8, 2015, to carry out restoration programs and projects for the Great Lakes, including toxic substances remediation, controlling invasive species and mitigating nonpoint source pollution.

Rep. Lois Capps introduced H.R. 1278, the "Water Infrastructure Resiliency and Sustainability Act of 2015," on March 4, 2015, to improve aging water infrastructure in local communities. Senator Benjamin Cardin introduced a companion bill in the Senate, S. 741, on March 16, 2015.

In May 2015 the EPA released a new rule on the definition of "waters of the United States" and the future enforcement of the act. Thirteen states sued, and on August 27 U.S. Chief District Judge Ralph R. Erickson issued an injunction blocking the regulation in those states. In a separate lawsuit, on October 9 a divided federal appeals court stayed the rule's application nationwide. Congress then passed a joint resolution under the Congressional Review Act simply overturning the WOTUS rule, but President Obama vetoed the measure.

References

- Water Pollution Control Foundation. "The Clean Water Act of 1987." Joan M. Kovalic et al. Alexandria, VA, 1987. ISBN 978-0-943244-40-2.

- "Primacy Enforcement Responsibility for Public Water Systems". Drinking Water Requirements for States and Public Water Systems. EPA. 2015-11-09.

- Shepherd, Katie. (27 May 2015). "Under new EPA rule, Clean Water Act protections will cover all active tributaries" Los Angeles Times. (Los Angeles) Retrieved 5 November 2015.

- Gershman, Jacob (9 October 2015). "Appeals Court Blocks EPA Water Rule Nationwide". The Wall Street Journal. Retrieved 22 October 2015.

- Mark Drajem and Katarzyna Klimasinska (1 February 2012). "EPA Shrinking 'Halliburton Loophole' Threatens Obama Gas Pledge". Bloomberg. Retrieved 22 March 2012.

- Bredickas, Vincent; Hartnett, Kim (1998-02-24). "Safe Drinking Water Act". Water Treatment Primer. Blacksburg, VA: Virginia Polytechnic Institute and State University. Retrieved 2010-03-21.

Permissions

Index

A

Abandonment, 38
Acid Mine Drainage, 7, 43, 55-63, 102
Acidification, 23-24, 60
Adaptation, 28
Agricultural Wastewater Treatment, 12, 95, 99, 109
Airline Water Supplies, 240
Amoebiasis, 192-194, 196-198
Animal Wastes, 12, 112-113, 246
Anti-degradation Policy, 245
Applicability, 71
Aquatic Toxicology, 231-233, 235-237
Atmospheric Pollution, 22

B

Batch Distillation, 174-176, 184, 186
Biochemical Oxidation, 97, 108
Brine Treatment, 103-104
Bromination, 190

C

Calcium Silicate Neutralization, 61
Carbonate Neutralization, 61
Chemical Oxidation, 37, 97-98, 157
Chemical Waste, 6, 43, 51-55, 63
Chloramination, 190
Clean Water Act, 3, 40, 51, 67, 127, 141, 234, 236, 243, 252, 256
Coagulation, 16, 148, 150, 170, 203
Cold Water, 7, 16, 173
Constructed Wetlands, 12-13, 41, 61-63, 89, 99, 112, 118, 125
Continuous Distillation, 174, 176

D

Demineralized Water, 158
Diagnosis, 14, 195-196, 202, 208, 210-213, 216, 223, 227-228
Dimensions, 72
Direct Discharge, 20
Disinfectants, 107, 114, 146, 201-202, 206, 238-239, 242
Disinfection, 6, 37, 40, 97-99, 114, 121, 124, 134, 138-139, 146, 152-155, 158, 160, 188-191, 203, 238-239, 241-242

Dissolved Air Flotation, 102, 107, 150
Distillation, 37, 96, 109, 133, 145, 147, 156-158, 170-187, 191
Drug Pollution, 6, 15, 32, 38-39
Dssam Model, 8, 67-68, 74-75

E

Ecological Effects, 15, 40
Effects on Ph, 58
Energy Requirement, 125
Eutrophication, 4, 19, 24-25, 40, 121
Exposure Systems, 232

F

Filtration, 9, 37, 89, 97-99, 103-104, 107, 114, 121, 130, 133, 137-138, 145, 148, 150-152, 159-166, 190, 197, 203-204, 206, 208, 212, 241
Firewater, 114
Flocculation, 104, 145, 148-150, 170, 203
Fractional Distillation, 172, 178, 183

G

General Improvements, 176
Groundwater Model, 3, 30, 68, 70-71
Groundwater Pollution, 3, 8, 15, 29-35, 37-38
Groundwater Remediation, 37, 157

H

Hydraulic Fracturing, 33-34, 133, 138, 241, 243
Hydraulics Capabilities, 81
Hydrological Inputs, 69
Hydrological Transport Model, 3, 30, 64, 66
Hydrology, 3, 7, 30, 65-67, 74, 76-77, 80-81, 85-86, 90-91

I

Indirect Potable Reuse (ipr), 135, 141
Industrial Wastewater Treatment, 11, 95, 99-100, 114, 116
Inorganic Chemicals, 238-239, 253
Integrated Hydrology/hydraulics, 85
Interactions With Surface Water, 35
Iodinization, 190
Ionizing Radiation, 191
Iron and Steel Industry, 101

L

Laboratory Waste Containers, 53

Lime Neutralization, 60

M

Manufacturing, 8, 11, 38-39, 45, 47, 101, 105, 107, 244

Marine Debris, 7, 18, 25-26, 43-45, 49-51

Marine Pollution, 15, 18-21, 29, 44, 49

Membrane Filtration, 98, 103, 133, 137, 152

Mitigation, 13, 22, 28, 109

N

Naturally Occurring, 4, 30, 32, 34, 43, 139, 189

Nitrate, 10, 30-32, 40, 42, 112, 122-123, 125, 130, 239

Nomenclature, 57

Nuclear Industry, 102

Nutrient Pollution, 12, 15, 19, 39-40

O

Operational Inputs, 69-70

Ozonation, 124, 137, 154, 190

P

Parameters, 30, 59, 64-65, 67, 69-71, 73, 75, 77, 79, 81-85, 87, 89, 91, 93, 234

Pathogens, 4-5, 10, 30-33, 37, 97, 124, 132, 136-137, 139, 153-154, 156, 159, 189-190, 200, 253

Pathways of Pollution, 20

Pesticides, 10, 12-13, 18-19, 27-28, 34, 100-101, 107, 109, 111-112, 114, 125, 156, 191, 239, 246, 253, 256

Piggery Waste, 113

Plastic Bags, 22, 25-26, 44-46

Plastic Debris, 8, 25-27, 44-45, 47-48

Polishing, 96, 98, 121, 127, 170

Potable Uses, 133-134, 140

Precipitation of Metal Sulfides, 62

Prediction, 57, 59, 64, 66, 75

Pretreatment, 115-116, 118, 138, 248, 254

Prevention, 11, 13, 20, 35, 39, 113, 131, 144, 153, 192, 196-197, 203, 213, 216, 218, 221, 228, 230, 250

Prognosis, 197, 213

R

Radionuclides, 32, 55, 239

Reclaimed Water, 95, 99, 131-132, 134-143

Reclamation Processes, 136

S

Safe Drinking Water Act, 37, 238, 241-243, 256

Sanitation Systems, 5, 30, 32, 35-36, 128, 194

Sedimentation, 89, 96, 102-104, 118-120, 127, 145, 148-150, 157, 166-170, 195, 197, 203

Sewage Mixing with Rainwater, 116

Sewage Treatment, 3, 10, 33, 38, 41, 95, 97-100, 107, 114-117, 121-122, 124-129, 135, 141, 161, 247-248

Sewage Treatment Plants, 10, 33, 98-99, 117, 122, 125, 127, 135, 141, 247-248

Ship Pollution, 21

Shock Chlorination, 189

Short Path Distillation, 180-181

Silage Liquor, 113-114

Sludge Treatment, 98, 119, 125-126, 253

Space Travel, 135

Storm Water Management Model, 65, 76-77, 91

T

Technology-based Standards, 244-245, 248

Thermal Pollution, 7, 14-17, 249

Thermal Shock, 15-16

Toxicological Effects, 232, 234

Toxins, 8, 19-20, 27, 121-122, 156, 191, 243

Treated Sewage Reuse, 128

Treatment, 1-6, 8, 10-12, 14, 16, 18, 20, 22, 24, 26, 28, 30, 32-34, 36-38, 40-42, 44, 46, 48, 50, 52, 54, 56, 58, 60-63, 65-66, 68, 70, 72, 74-78, 80, 82-84, 86, 88, 90, 92, 94-110, 112-144, 146-152, 154-162, 164, 166, 168, 170, 172, 174, 176, 178, 180, 182, 184, 186, 188, 190-192, 194, 196-198, 200, 202-206, 208, 210, 212-216, 218-222, 224-226, 228-230, 232, 234, 236, 240, 242-244, 246-248, 250, 252-254, 256

Treatment Of Other Organics, 109

Treatment Of Toxic Materials, 109

U

Uv Radiation, 124, 155, 191

V

Vacuum Distillation, 178-181, 186

Volatile Organic Compounds, 6, 11, 31, 241

W

Warm Water, 16

Waste Water, 122-124, 137, 141

Wastewater Treatment, 10-12, 30, 75, 95-100, 109, 113-116, 121, 125-126, 128, 133, 144, 170, 234, 243, 253-254

Wastewater Treatment Plants, 30, 95, 98-99, 121, 128, 133, 234

Water Chlorination, 145, 160, 187-189

Water Purification, 37, 134, 145, 147-149, 151, 153, 155-157, 159-161, 163, 165, 167, 169, 171, 173, 175, 177, 179, 181, 183, 185, 187-189, 191

Water Quality Criteria, 236, 245, 248

Water Quality Standards, 40, 231, 236, 244-246, 248-250, 255

Wool Processing, 103